Psychologies of Ageing

Elizabeth Peel • Carol Holland
Michael Murray
Editors

Psychologies of Ageing

Theory, Research and Practice

palgrave
macmillan

Editors
Elizabeth Peel
School of Social Sciences
Loughborough University
Loughborough, UK

Carol Holland
School of Health and Medicine
Lancaster University
Lancaster, UK

Michael Murray
School of Psychology
Keele University
Newcastle-under-Lyme, UK

ISBN 978-3-030-07288-9 ISBN 978-3-319-97034-9 (eBook)
https://doi.org/10.1007/978-3-319-97034-9

Cover illustration: Jayesh/gettyimages

This Palgrave Macmillan imprint is published by the registered company Springer Nature Switzerland AG
The registered company address is: Gewerbestrasse 11, 6330 Cham, Switzerland

Foreword

I feel honoured to be invited to write this Foreword, to a positive presentation of the psychologies of ageing and of dementia. So much has changed since 1995, when I was diagnosed with dementia. I was only 46 and a recently divorced single mother of three girls aged 9, 13, and 19. We had just moved at the weekend into our new house and were looking forward to a bright new future. However, on the Monday after our move, I had a follow-up appointment with a neurologist, taking an hour off from work to do this. I had been suffering from extreme headaches each week, and what I thought were symptoms of stress, such as occasionally becoming confused with words, as well as with finding my way on familiar routes. There had been a few months of scans and tests, and now I was looking forward to finding out what was wrong.

With his back towards me, while looking at my scans and test results, the specialist calmly said that my symptoms of what I thought were stress were in fact due to dementia. I exclaimed: "How could that possibly be, as this is something that affects much older people!" Then, with an impassive face, the neurologist told me he had diagnosed quite a few younger people, even a pregnant woman in her early 30s. He told me to get my affairs in order, as soon as possible, as I might not even be able to sign my name by the end of the year. There was no treatment and no hope for the future.

Coming out of his room, I was reeling with shock, and driving back to work was a nightmare: what would happen to my girls; could my 19-year-old look after my affairs? Questions swirled around in my head, and it was very hard to concentrate for the rest of that day, let alone the week. I took some much-needed leave in order to seek a second opinion, as well as to begin to deal with the implications of this truly awful diagnosis. Diagnosis with dementia at the age of 46 was earth shattering and quickly led to my isolation from friends and former work colleagues, as I became withdrawn, depressed, and unable to cope.

Where would I find psychological support? When I rang the Alzheimer's Association, I was asked for whom I was caring. When I said that I needed this support for me, I was told there was nothing for the person with dementia, only for older carers, facing an awful journey ahead. However, what about the person with dementia and their own grief and loss? What about younger carers? My eldest daughter was only 19 and away at university. There was no support for me or for my young carers. As a family, we did not fit into the usual picture for dementia.

Back then, in the mid-1990s, people with dementia were thought to be unable to speak and lack insight and to be around 80 or 90 years old. The common view was that dementia is a normal part of ageing and not the consequence of a range of diseases that could occur even at a young age. Support for people with dementia was not thought to be feasible, nor even necessary, as we lacked insight and were unable to speak. Indeed, if you could speak or have insight, the diagnosis was questioned. So I was left struggling without support for my girls or me and became increasingly lonely and isolated. The question that was foremost in my mind was: Who will I be when I die? I had been told that I would decline, probably rapidly for about five years, and then go into full-time care, until I died within about eight years. I believed this dismal prognosis, so my function did in fact become more impaired due to depression; yet there was no support to help me to cope during this time.

It has been a long and lonely struggle after my diagnosis, surrounded by negative assumptions about dementia. I began a journey of writing and speaking about dementia from an insider's perspective, trying to provide an alternative discourse. I wanted to highlight the need for ongoing psychological support, beyond simply the clinical testing that assesses our functional abilities. I argued for social support and empathy, and an acknowledgement of our continuing identity, despite negative views about a so-called future loss of personhood, when it seems as if our identity would become dependent on others to sustain.

I wanted the community to recognise that people with dementia, even at a younger age, should have access to appropriate in-home and residential care, which gives positive support. Importantly, we need recognition of our feelings of isolation, when we can no longer drive. Not only do we lose our sense of independence but also become increasingly reliant on public transport, which could well become confusing. When I looked into the options for residential care, I realised that much older people would surround me, with therapy designed for this age group. Again, the assumptions were that dementia was a normal part of ageing and not a condition of younger people.

Importantly, only recently has there been any information about what can be done in terms of lifestyle changes to improve our function, or indeed for the prevention of dementia. After diagnosis, I faced a bleak picture of future decline with no hope, where there was nothing that could be done. Indeed, even the anti-dementia medications were not thought to be useful, so were often not prescribed. This lack of future hope drove my efforts to change the narrative towards more positive views of people with dementia and increasing efforts to provide a range of supports.

However, I have outlasted the prognosis of a decline until death with dementia within 8 years, as it has now been over 20 years since that fateful day, which dramatically changed my life story. As I often say, "I'm still here!" Now I am almost 70, and perhaps it is finally more appropriate to have a diagnosis of dementia. Looking to the future, this will no doubt be a time of reflection on my past, as well as the years that might still lie ahead.

Now I face both ageing and dementia, which present me with the "double whammy" of negative social perceptions, where both social groups are thought to be a costly burden to society. Rarely do we hear the counter-story of our having contributed to the society that we enjoy today, nor indeed any reference to the wisdom we might have accumulated over a lifetime of achievement. Yet the aged, as well as people with dementia, are just as diverse as we once were, still worthy of support and inclusion, and of being given dignity and respect.

However, increasing age has brought with it another issue: that of frailty. A few years ago, I had a fall and broke my hip, arm, and cheek. It took around a year or so to recover, but even then I was still cautious and uncertain. I felt frail and fearful of further falls. It took quite a few years for me to regain my confidence and the resilience that I needed to cope with the possibility of falling once more. Again, just as had been my experience with dementia, I found there was little support available to overcome frailty and my fear of falling. However, finally now I am able to exercise again without fear, although I always take my mobile phone with me just in case.

If only this book had been available many years ago! It provides an excellent and comprehensive overview of many of the concerns that have troubled me over the years and would have overcome many of the issues that I faced as a person with dementia, who was diagnosed at the age of only 46.

Now, as I become older and frail, the book counters the negative social views of ageing, where even the political narrative is changing. Challenging negative views has an enormous, often under-recognised impact on the individual with dementia and/or who is ageing. As a person who has struggled to overcome negativity for over 20 years, I welcome this wonderful book.

Charles Sturt University
Sydney, Australia

Christine Bryden

Acknowledgements

We would like to thank all the contributors for their excellent chapters, Dr Christine Bryden for providing such a compelling personal foreword, and Grace Jackson and Jo O'Neill at Palgrave Macmillan. The British Psychological Society funded the research seminar series *Beyond Boundaries: Exploring Psychologies of Ageing* (2015–2016) from which this volume arises, and further discussion of this collaborative project can be found in *The Psychologist* and *Working with Older People*.

Contents

Notes on Contributors

Mary Breheny is Senior Lecturer in Public Health at Massey University, New Zealand, and member of the Health and Ageing Research Team. Her research focuses on the ways that inequalities throughout the lifespan accumulate in later life and constrain older people from ageing well.

Sarah Canham is a university research associate with the Gerontology Research Centre and a senior lecturer with the Department of Gerontology at Simon Fraser University, Canada. Her interdisciplinary research has revolved around topics of mental health and ageing and the social contexts of substance use, loneliness, homelessness, and service delivery. She also has an active role in the Gerontological Society of America, where she is the convener for the *Aging, Alcohol, and Addictions Interest Group*.

Jordan Elliott-King is a PhD student at Loughborough University, UK. She has worked with people with dementia for three years and has written on the topic of behavioural and psychological symptoms of dementia, dementia diagnostics, and verbal memory.

Mei Lan Fang is a transdisciplinary research scientist with a passion for understanding and exposing, across societies, the sociocultural and structural foundations that shape individuals' experiences of marginalisation throughout the life course. She is a PhD candidate at Heriot-Watt University, Scotland, where her doctoral work focuses on how the experiences and histories of older Canadian migrants impact their transition into affordable housing in Metro Vancouver. She is also a senior scientist in Knowledge Translation at MTEK Sciences, a

Vancouver-based global health research consulting firm funded by the Bill and Melinda Gates Foundation.

Joseph Ford is a postdoctoral research associate at the University of Exeter Medical School, UK, where he analyses primary care consultations about mental health. He completed his PhD on Empathy in Palliative Care from the Department of Social Sciences at Loughborough University, where he also acted as a co-ordinator of the Discourse and Rhetoric Group (DARG) from 2015 to 2016.

Ian Garner is a PhD research student based within the Aston Research Centre for Healthy Ageing (ARCHA), UK. ARCHA uses a multidisciplinary approach to successful ageing by researching technological, therapeutic, and psychosocial strategies as a means of combatting age-related decline. Garner completed his undergraduate degree in Psychology and postgraduate degree in Family and Child Psychology before continuing to specialise in resilience research. Prior to his position in ARCHA, he spent two years conducting resilience research in primary schools, implementing a cross-age teaching scheme (CATS) as a method of promoting resilience to combat the negative psychological impact of long-term and severe bullying. His work focuses on the development of a resilience-building intervention to be implemented in supported living environments with the ExtraCare Charitable Trust.

Holly Gwyther is a Chartered Psychologist with a multidisciplinary background in psychology and biology. Her research interests include mixed-methods research with vulnerable groups including older adults and women. Specific interests include the effects of ageing on cognition and wellbeing and risk behaviours examined through health psychology models. Gwyther has expertise in critical analysis, evidence synthesis, and multiple methods research. Her EU-funded work focuses on optimising management of frailty in older adults across Europe (FOCUS), based within the Aston Research Centre for Healthy Ageing (ARCHA).

Sharron Hinchliff is Reader in Psychology and Health in the School of Nursing and Midwifery, University of Sheffield, UK. She has been conducting research into gendered health for over 20 years. Hinchliff leads the ageing, gender, sexual health, and sexual wellbeing programme of research at the University, which examines the psychological, social, and cultural factors that influence health and healthcare. Her research is concerned with changing social attitudes towards older people, sex, and relationships, and it aims to improve professional healthcare practice in this area. She has written extensively for academic, clinical, and lay audiences, and her latest book *Addressing the Sexual Rights of Older People: Theory, Practice and Policy* was published in 2018.

Eef Hogervorst is Professor of Biological Psychology at Loughborough University, UK. She has researched dementia for over 20 years and worked at both Oxford and Cambridge University. She has written over 200 peer-reviewed publications on this topic and obtained over £9 million research money to do her work with collaborators. Hogervorst reviews for many journals and funding bodies and is regularly invited to speak about her research to international scientific communities and to the lay public.

Carol Holland is Professor of Ageing at Lancaster University, UK, and Director of the Centre for Ageing Research and previously Director of the Aston Research Centre on Healthy Ageing. She is the principal investigator for several well-known studies such as the evaluation of extra care housing for older adults and UK Principal Investigator for the FOCUS project on frailty, and is a co-ordinator for the European Innovation Partnership on Active and Healthy Ageing. She is a Chartered Psychologist and associate fellow of the British Psychological Society.

Andrew King is Professor of Sociology at the University of Surrey, UK. He has been undertaking sexualities research for over 15 years and researching Lesbian, Gay, Bisexual, and Transgender (LGBT) ageing for over 10 years. His LGBT ageing research has been funded by, amongst others, the Economic and Social Research Council (ESRC) and the EU. He has addressed the relationship between older LGBT people and service providers, particularly in the area of housing. He has written widely on LGBT ageing, in a range of edited collections and articles in *Sociology*, *Ageing and Society*, *International Social Work*, *Journal of Sociology*, and *Social Policy and Society*. His book *Older Lesbian, Gay and Bisexual Adults: Identities, Intersections and Institutions* was published in 2016 and he is the co-editor of *Sexualities Research: Critical Interjections, Diverse Methodologies, and Practical Applications* (2017).

Daniele Magistro is a Lecturer in the Sports Science Department at Nottingham Trent University, UK. Magistro's research priority is understanding, preventing, delaying, and/or avoiding risks related to ageing. He focuses on new methodologies and training to gain insight into the amount, intensity, quality, and types of physical, cognitive, and daily activities required for optimal health and an independent life in older adults.

Michael Murray is Professor of Social and Health Psychology at Keele University, UK. He has written over 100 journal articles and chapters on a range of issues in social and health psychology including ageing, community health, and occupational health. He has also written theoretical and methodological chapters on narrative psychology, community health psychology, and social representation theory. He has co-authored *Health Psychology: Theory, Research and Practice* (five editions) and co-edited *Critical Health Psychology* and *Qualitative Health Psychology: Theories and Methods* (with Kerry Chamberlain). Much of his current research uses participatory approaches designed to encourage community transformative action to promote health and wellbeing among older people.

Charles Musselwhite is Associate Professor of Gerontology at the Centre for Innovative Ageing (CIA), Swansea University, UK, and leads the Environments and Ageing research strand. He has particular interests in ageing, travel, and transport, addressing technological, environmental, health, and sustainability contexts of transportation and built environment studies.

Elizabeth Peel is Professor of Communication and Social Interaction and Associate Pro Vice-Chancellor (Doctoral College) at Loughborough University, UK. She is a fellow of the British Psychological Society; a critical social psychologist with interests in gender, sexualities, and chronic illness in later life. Her most recent books are *Ageing and Sexualities* (with Rosie Harding) and *Critical Kinship Studies* (with Damien W. Riggs). She held an inaugural British Academy Mid-Career Fellowship for the Dementia Talking: Care, Conversation and Communication project.

Judith Sixsmith is Professor of Health-Related Research in the School of Nursing and Health Sciences at the University of Dundee, Scotland. Prior to that, she was Professor of Public Health Improvement and Director of the Institute of Health and Wellbeing at the University of Northampton. She is also a Visiting and Adjunct Professor of Public Policy at Simon Fraser University, Vancouver, Canada. Her research interests lie in the areas of public health and social care where she explores the ways in which people living in disadvantaged communities experience processes of marginalisation within our social systems.

Christine Stephens is Professor of Social Science Research at Massey University, New Zealand. She teaches in critical health psychology and co-leads the Health and Ageing Research Team to conduct longitudinal and qualitative research on the health of older people. She has a particular interest in the contribution of social participation to wellbeing.

Anna C. Whittaker (previously Phillips) is Professor of Behavioural Medicine in the School of Sport, Exercise & Rehabilitation Sciences, University of Birmingham, UK. She is an internationally renowned researcher and Registered Health Psychologist working in Psychoneuroimmunology and Psychophysiology, particularly in the context of ageing. She has conducted award-winning work on stress and vaccination response across the life course and has a strong profile of public engagement with science. She is also now developing her research into the area of physical activity interventions for older adults to improve health, function, and wellbeing through leading the European Commission Physical Activity and Nutrition INfluences In ageing (PANINI) training network and consortium across Europe.

Katie Wright-Bevans is Lecturer in Social and Community Psychology at Keele University, UK. Her research focuses broadly on community wellbeing and social inclusion using critical social psychological theories and participatory qualitative methods. Wright-Bevans has a particular interest in action research with both marginalised and intergenerational groups as a tool for understanding stigma, marginalisation, and sources of inequality, as well as identifying avenues of resistance and change, and she has been involved in numerous collaborative research projects with community groups across the North of England. Wright-Bevans's PhD thesis critically examined intergenerational practice as a means of promoting community cohesion, wellbeing, and social inclusion. She is co-chair of the Keele Action Research Network and is a member of the British Psychology Society Community Psychology section committee.

List of Figures

List of Tables

1

The Psychologies of Ageing

Michael Murray, Carol Holland, and Elizabeth Peel

Psychology contains a wide diversity of approaches from the biological through to the societal. Each of these approaches makes a contribution to our understanding of ageing. The aim of this edited collection is to highlight the contribution of a series of key approaches which provide a unique perspective on ageing. These perspectives were chosen to illustrate current psychological research on ageing which adopts a more critical and less reductionist approach. As an introduction to this collection, this chapter considers briefly the tremendous challenge and

M. Murray (✉)
School of Psychology, Keele University, Newcastle-under-Lyme, UK
e-mail: m.murray@keele.ac.uk

C. Holland
School of Health and Medicine, Lancaster University, Lancaster, UK
e-mail: c.a.holland@lancaster.ac.uk

E. Peel
School of Social Sciences, Loughborough University, Loughborough, UK
e-mail: e.peel@lboro.ac.uk

E. Peel et al. (eds.), *Psychologies of Ageing*,
https://doi.org/10.1007/978-3-319-97034-9_1

opportunities posed by our ageing society and then reviews the main psychological perspectives considered and how these have engaged with this phenomenon.

Ageing

Growing old is part of life. However, what it means to grow old in the twenty-first century is dramatically different than in previous eras. Our understanding of what it means to age, and 'age well', has shifted in recent decades, with some scholars highlighting the obsessional heights that this has reached (Lamb 2017). People throughout the world are living longer than ever before and the proportion of the world's population who are classified as old continues to increase dramatically. According to the World Health Organisation (2015), the proportion of people in the world who are over 60 years of age will increase from 12% to 22% by 2050. In actual figures, the number of older people will increase from 900 million in 2015 to 2 billion people in 2050. In addition, people will be living longer. The WHO estimates that the number aged over 80 years will increase from 120 million in 2015 to 434 million in 2050. Further, by 2020, the total number of people aged 60+ will be greater than those aged less than 5 years by 2020. This rapid increase in the world's older population brings many associated challenges as well as opportunities.

The growth of the ageing population has led to sustained discussions by national and international agencies on how to wholly ensure that older people have the fullest opportunity to participate in society and that the various support agencies are designed to address their needs. The World Health Organisation (2007) has been to the fore in developing a wide range of initiatives. One of the most important initiatives has been the promotion of age-friendly cities and communities. Sustained research and consultation with older people and various support agencies identified various domains as important. While these domains are key to the development of healthy ageing, there is a need for more social and psychological research designed to explore in more detail their content.

Balancing the opportunities with responding to the challenges is a focus that underlies much of this collection that examines the different roles psychology plays in the various spectra of ageing. Throughout the ageing literature in the last 40 years we have seen various iterations of terms such as 'successful ageing' (see Bowling and Dieppe 2005, for a review of this term), 'active ageing', 'healthy ageing', 'positive ageing', or 'optimal ageing', each bringing its own perspectives and both positive and negative connotations, which can be seen in the different themes throughout the book. It is worth spending a little time thinking about each of these terms as readers see different viewpoints on them throughout this book.

When we consider how ageing is typically represented within psychology, there is an implicit, and sometimes explicit, focus on individual and intrapsychic contributing factors. To take Ian Stuart-Hamilton's (2012) popular, and comprehensive, textbook on *The Psychology of Ageing* as an illustration of this approach; of the nine chapters most foreground cognitive processes or individual differences. While not wishing to critique this focus it is telling that there is a chapter on measuring intellectual change in later life, another on ageing and memory, and also a focus on ageing and language in which the material largely covers linguistic constraints and cognitive processing issues. Stuart-Hamilton does not take an uncritical stance to the 'confounds' within the mainstay of psychological science on ageing but quite the reverse. Nevertheless, the underpinning psychological research on which his textbook, and arguably the field, is based is skewed towards more individualistic accounts of ageing. As a counterpoint, in constructing this collection, we aim to explicitly label and illustrate different forms of psychological research on ageing which move beyond the traditional canon within the field. The goal being to allow the reader to make more visible links between the underpinning psychological approach and the research questions and findings that flow from that approach. This form of psychological research has clear practical implications to challenge those negative social constructions and institutions which exclude older people from full participation in the social world.

Psychological Theories

Critical Social Psychology

Critical social psychology and analyses of discourse have made a growing contribution to the psychologies of ageing. This paradigm, broadly conceived, can reveal the social construction of ageing and offer analytic tools to examine, and also challenge, ageism. In part, as a counter to cultural understandings of ageing as synonymous with frailty and decline, the term 'successful ageing', which was first coined in gerontology in the 1960s but has grown in prominence since from the 1980s onwards (Lamb 2017), now forms a key framework in understanding the psychology of ageing. Perceptions of conditions like dementia and frailty as being a normative part of ageing—just things that happen with increasing age—has now almost gone because of these changes, with the resulting impact on research and the fact that now more is understood about risk factors for dementia and frailty for instance. A positive aspect of successful ageing is that it supports arguments for better support and funding for preventive rather than crisis only based health and social care.

Probably the most commonly used definition of successful ageing is satisfaction with one's past and present life (Havighurst 1963), but other more biomedical definitions have also had an influence. These views conceptualise successful ageing more in terms of the optimisation of life expectancy while minimising the gap between longevity and healthy life expectancy, with Rowe and Kahn's (1998) definition also having significant influence. This definition included the absence or avoidance of disease and risk factors for disease, maintenance of physical and cognitive functioning, and maintenance of social engagement and involvement.

While the number of definitions have accumulated, some have suggested that the concept of successful ageing is informed by and grows out of neoliberal notions of individual responsibility for how we age, and, according to Lamb et al. (2017) 'has persisted as the dominant paradigm in gerontological research, appearing also under such related labels as "active aging", "healthy aging", "productive aging", "vital aging", "anti-aging", and "aging well"' (p. 1). Critical social psychology and discursive approaches which are considered in Chapters 2, 3, and 4 offer insights into how such new discursive, and material, landscapes shape the experiences of older people and indeed critique models such as successful ageing.

Stephens and Breheny, in Chapter 2, take a social constructionist approach highlighting how individualisation and healthism work to 'defy ageing itself'. They demonstrate how the medical discourse of ageing, and successful ageing, is rooted in social policy discourse and how individual responsibility discourse interleaves to construct the older 'citizen who is responsible for their own physical and financial wellbeing'. Stephens and Breheny draw two contradictory discourses which exemplify the contradictory discursive landscape older people occupy from their interviews with older New Zealanders, namely 'productive time' and 'personal time'. In the former, leisure activity is useful and demonstrates the achievement of a productive, active citizen; in the latter the emphasis is on personal choice and freedom justified through an emphasis on contributing throughout the life course. Ultimately, and importantly, they demonstrate how 'language shapes research, practice, and policy'; this, along with the 'how' provided by conversation analytic methods as demonstrated by Ford, in Chapter 4, signals the value and significance of critical social and discursive approaches to the psychology of ageing.

Discursive approaches to ageing take many forms. At the level of interrogating participants' accounts, age appropriateness and norms around ageing can be illuminated by, for example, understanding how contradictory accounts of 'old' and 'a little girl' are produced in talk (Nikander 2000). In Chapter 3, Hinchliff and King allude to the utility of membership categorisation analysis as an approach to understanding older lesbian, gay, and bisexual (LGB) peoples' use of identity categories within specific instances of talk. They then deploy intersectional narrative analysis to explore some of the ways that gender, social class, religion, and ethnicity intersect in older LGB peoples' accounts. Moreover, they move outside the conventional parameters of the discipline—a vital critical move—to bring sociology and intersectional analysis into conversation with social psychology on the topic of ageing and sexualities (Peel and Harding 2016).

In Chapter 4, Ford offers a thorough exploration of how the 'discursive turn' in social psychology provides an important framework for engaging with psychological processes not as cognitive phenomena but as interactional achievements. Taking displays of empathy in palliative care interactions as his focus, and using conversation analysis to scrutinise how doctors create interactional space for older patients to talk about their condition, or acknowledge emotional and experiential difficulties caused by physical symptoms, Ford shows how a person-centred approach in later life can be enacted.

Biological, Cognitive, and Health Psychology

Many of the terms used to describe ageing come from an understanding that variability in a range of factors increases, the older the population sample being studied, and that health challenges begin to accumulate, but not for everyone. For example, although the number of diagnoses of chronic diseases a person may have increases with age, to the extent that 81.5% of over 85-year-olds have two or more chronic conditions, that still indicates that 18.5% of this age group have no chronic diagnoses, or only one (Salive 2013). Likewise, although the likelihood of cognitive change increases with age, in any study of cognitive function, there are people in the oldest groups who show no decline, with population distributions of performance showing fewer people clustered around the mean and longer 'tails' of people performing at greater distances from the average, the older the population group. Within cognitive functions, measures of variability increase with increasing age (e.g. Morse 1993) and within-person variability (either within a task from moment to moment, or between different functions) have both been shown to be indicative or predictive of serious impairment, including later Alzheimer's disease (Lövdén et al. 2007; Hilborn et al. 2009; Halliday et al. 2018).

Both the biological and cognitive psychology of ageing have this understanding of variability at their base. If one 80-year-old has health and cognition that is very different from another 80-year-old, or if each has some aspects of good health or cognition and some poor ones, then we must conclude that chronological age, in itself, is not causing the changes. This leads us to look for the underlying reasons for observed changes with a view to identifying risk factors, mechanisms, and potential modifiers and so interventions. The interaction between our physiologies, brain, behaviour, and our environment, including our social situation, is a key focus for psychologists and this is highlighted in the chapters in this section.

Biological psychology explores the biological bases of our behaviour and can also be called physiological psychology. The sub-discipline focuses on the interaction between our physiology, including the brain, and our behaviour. While the focus is usually on internal processes and the field can be described as reductionist, the interaction with the environment is crucial. This is typified in Chapters 5 and 6. Chapter 5 discusses the impact of stress coming from the environment on physical health via the body's physiological responses to stress, and so to its impact on the immune system. Stress can be defined as being in a situation that exceeds our coping abilities, and we have been aware for some time that people undergoing stress are more susceptible to a range of disease processes. Whittaker, Chapter 5, offers some explanations for these links, first discussing some of the mechanisms by which stressful situations and our experience of coping with them have an impact on our immune system. Central to the development of the work in ageing immune systems is the concept of a balance between inflammation and immune system factors necessary to deal with it, which can change with increasing age, known as inflammaging. The importance of this balance in longevity, in the development of chronic diseases, cancers, dementia (Giunta et al. 2008), and frailty (Feng et al. 2017) is only now beginning to be understood.

Whittaker (Phillips) describes a series of significant articles on the impact of stress on health, notably examining the impact of caring for a child with developmental disorders or a family member with dementia, demonstrating that, for example, the response to vaccination against influenza is poorer in caregivers, particularly older carers. Responding to the question of poorer health following stressful life events, Whittaker describes a study examining the impact of stressful life events on the response to the influenza vaccine amongst younger and older people. While finding fewer differences than expected, the older participants did show an effect of having experienced bereavement in the last year

(Phillips et al. 2006). This mirrors other research that has directly examined immune function in bereaved younger and older adults, showing an effect only for the older people (Vitlic et al. 2014).

Indeed, immune function has been suggested as one mechanism whereby cumulative models of frailty, such as that described in Chapter 7, may have their impact. Whittaker introduces studies on the impact of developing depression on likelihood of infection following hip fracture and of the impact of social support and social networks on immune function, underlining the need to take a broader view of biological psychology in the context of older age. Indeed, the role of strong social support, for example, from happy marriages, was suggested to be a useful mechanism that counteracted the effects of ageing on immune functioning and that may have a direct impact rather than simply an impact on reducing stress. As with all the chapters in this section, the issues are presented as those which can have an impact via our behaviour, and Whittaker suggests health behaviours effect both stress and immune function, with a recent article demonstrating the role of physical activity in reducing the impact of older age on the immune system (Duggal et al. 2018).

The theme of health behaviour is continued in Chapter 6, which also introduces the impact of biological underpinnings on cognitive function in older age. Here the interaction between the impacts of health behaviour on one's brain health and the risks of cognitive decline and dementia in older age is examined. The evidence behind the relationships between a physically active lifestyle and reduced risk of cognitive decline in older age is examined using longitudinal studies, alongside studies that take a normal range of older adults and randomly allocate them to exercise groups and controls, both demonstrating a positive impact of lifetime activity and new interventions in older age. Importantly, this has been shown to have an effect in people who already have a diagnosis of Mild Cognitive Impairment (MCI) (e.g. Larson et al. 2006) and in people with the main genetic risk marker for Alzheimer's disease, APOE-e4 (Head et al. 2012).

The roles of lifetime and later life cognitive activity are also considered, again using longitudinal studies and later life interventions. Elliott-King and her colleagues (Chapter 6) and Holland and her colleagues (Chapter 7) suggest that group activities are often more suitable for older adults.

Group activities may help with strategies for adherence, but importantly, they also provide a sociable and fun aspect, often improving a person's perceived social support. Both Chapters 6 and 7 specifically concern themselves with the role of health psychology in terms of understanding why people make positive lifestyle choices and why some people do not, or are unable to. While understanding of the benefits, mechanisms and likely impact of exercise, healthy diet, an intellectually engaged lifestyle, or taking steps to reduce the chances of social isolation is a crucial role for psychologists, so too is work on enabling people to live healthier lives via work on developing health literacy and interventions that work in different contexts.

Chapter 7 takes a mainly qualitative approach to understanding the experiences, beliefs, and perceptions of a range of actors including the frail and not frail older person, their carers, health and social care practitioners, and senior policymakers in relation to interventions. This approach, together with theoretical underpinnings using more quantitative models of behaviour change, demonstrates the usefulness of a mixed-methods approach in such challenges and ensures that the perspectives of older people and those who care for them are an integral part of any programmes to improve health outcomes. This goes further than the individual, leading us back to consider the full socio-political, environmental, and organisational contexts we all find ourselves in.

Finally, in this section, the concept of resilience is considered in depth. The notions of healthy ageing or successful ageing that this book considers throughout can give a misleading impression that anything less than perfect health and maintained activity and involvement at mid-life levels is not healthy ageing, potentially leading to negative perceptions about ageing in any other way. However, the study of resilience as an important component of health that is independent of physical ailments or fitness leads us to consider the concept both of positive ageing and of health, whereby health is a state of balance. The WHO defines health as 'a state of complete physical, mental and social well-being and not merely the absence of disease or infirmity' whereas Richardson et al. (2011) comment that a sense of wellbeing and social acceptance is crucial, but does not necessarily mean being free from disease, disability, or frailty. Indeed, concepts of control and autonomy feature highly in people's descriptions of health and wellbeing, with frail older adults in Ebrahimi et al.'s (2013) study emphasising 'the importance of maintaining consistency and predictability in biopsychosocial contexts; having a manageable everyday life creates a sense of assurance and strengthens the older adults' experience of health' (p. 291). The overriding theme that emerged from their investigation of frail older adults' experiences of health offers a useful summary to the second section of this collection whereby psychological responses, self-efficacy, and control have been underlying themes to the connections between brain, body, and behaviour.

Societal Psychology

Not only psychology but other scientific disciplines have expanded their research into ageing. Walker (2014) in his review of the development of the New Dynamics of Ageing research programme noted that a common focus of this expanding research agenda has been a concern with the person-environment perspective. The key element of this perspective has been consideration of the social, economic, physical, and cultural/political context within which ageing occurs. This perspective has promoted much contemporary multi-disciplinary research on ageing which has encouraged many psychologists to work alongside other social and health scientists. These psychologists have been informed by a multi-disciplinary perspective, drawing upon ideas from neighbouring disciplines to develop a more sophisticated understanding of ageing. This approach accords with the call by Hilde Himmelweit (1990) for a societal psychology. By this she meant a psychology that considers the social, political, cultural, and environmental contexts within which we as humans are located. This is not to ignore the intrapsychic and interpersonal processes but rather to locate them within a broader societal context.

In the third section, we consider community, environmental, and socio-cultural psychological perspectives on ageing. Each of these approaches overlaps with the others such that it is difficult to talk about community interventions without considering the broader cultural and political context. Similarly, concern with environmental factors needs to be located within the cultural and political context. This expansive approach connects with the critical social psychological approach detailed in the first section of this collection in its aim to both critique and to challenge those societal arrangements that oppress older people. It is focused on working with older people to enhance their everyday lives.

Community psychology entails understanding how people connect with their communities as well as working with communities to promote social change. It developed in North America from the community mental health movement in the 1960s which evolved as part of the push for de-institutionalisation of care for people with mental health issues. In Latin America, it had a more political origin during the 1960s and was concerned with working with communities to defend them against various forms of social oppression and to promote efforts for broader societal change. Over the past 50 years the approach has evolved into promoting participatory forms of collective action and social change to promote social wellbeing. Increasingly it has taken up broader political issues and explored how to build alliances for social change (Murray and Wright-Bevans 2017; Nelson and Prilleltensky 2010). Further, the definition of community has expanded from one based upon local physical connections to an emphasis on collectivities in general.

The World Health Organisation (2002) identified both housing and outdoor spaces and buildings as two of the eight domains of an age-friendly community. In Chapter 8, Sixsmith and colleagues use participatory forms of research to explore how older people engage with these two key domains. Older people define themselves with reference to these neighbourhoods. They are likely to have spent the majority of their lives in the same home or neighbourhood and to have developed a sense of place which provides them with a certain rootedness. They locate themselves within their memories of that place.

In a study of the experiences of a disadvantaged urban community, Murray and Crummett (2018) reported that the older residents identified strongly with the immediate locality. It was part of who they were despite the many adversities they faced. Many of the residents had lived in the same houses for almost 40 years. They had raised their children there and formed close relationships with their neighbours. Their sense of community was reflected in the stories they told about their neighbourhood. They recalled the good times and the various hardships they had experienced. Despite their positive recollection of their community, they were frustrated by the negative outsider representation.

A key issue was the lack of control they had over the planning of their community. A frequent complaint was that the council officials ignored them. The work of Langer and Rodin (1976) confirmed the importance of perceived control in the lives of older people. They conducted an experimental study in which the residents of a nursing home were randomly assigned to either a group were they were told they had control over aspects of their everyday lives and another group who did not have such control. Eighteen months later it was found that those residents in the intervention group had improved health whereas the death rate was higher among the other residents. Mallers et al. (2013) identified this study as a classic in the field of gerontology. According to them it showed that it was better to allow older adults to do things for themselves rather than have them done for them. Subsequent research has confirmed that both perceived and actual control have positive influences. Even with frail older adults the benefits of minimal control have been demonstrated. For example, Lidz et al. (1992) found that older adults who, given the ability to make choices, understand options and have goals supported had improved wellbeing. This was the case even if assistance was required.

Sixsmith and her colleagues (Chapter 8) address this issue of control over their environment through a collaborative project with older people. Their aim was to work with the older people to co-create meaningful housing solutions transitioning into affordable housing. Central to their research was the importance of building community partnerships. Their research confirmed that although the process of building community partnerships was laborious, it was a necessary process in enabling older people to live independently and age well within their homes and communities.

Environmental psychology evolved from a focus on exploring the impact of the physical environment on human behaviour to a much broader concern with the relationship between humans and the natural, built, and social environment. In Chapter 9, Musselwhite considers one aspect of the environment—the use of transport. Again, the WHO (2007) identified transportation as one of the eight domains of an age-friendly community.

Any study of older people's social needs frequently brings to the fore discussion about accessibility. It is insufficient to develop new resources for older people unless these are easily accessible. For example, Newman et al. (2013) explored the value of access to art galleries for older people. However, when these older people were interviewed it was found that while they agreed that cultural engagement was important for their well-being they also emphasised the importance of transport to those facilities. Similarly, Murray and Crummett (2018) found that the older residents they worked with identified the lack of transport as the major deterrent to them accessing facilities outside their neighbourhood. Musselwhite, in Chapter 9, considers the various uses of transport by older people. He emphasises that transport must be understood within its societal context and with reference to the various social and psychological needs of older people. Like community psychology, environmental psychology emphasises developing approaches that will enhance everyday lives of older people. It is also involved in critiquing those broad environmental constraints on older people's lives whether it be in the design of buildings or the various social opportunities to which old people have access.

Socio-cultural psychology is an expansive approach which seeks to understand our everyday interaction with the social and symbolic world. As such it connects with the domains of the age-friendly city concerned with respect and social inclusion, social and civic participation, communication, and social support (WHO 2007). It includes within it a range of different approaches and methods from the descriptive survey-based approaches to the more interpretive approaches. The final chapter by Wright-Bevans and Murray connects with the more critical interpretive approach by exploring how old age is socially represented in society and how it can be challenged.

Social representations of ageing are constructed and circulated in everyday social interaction. They guide our everyday social expectations and interactions. However, certain agencies have greater power to shape dominant representations which are conveyed by the media and various social institutions. In the final chapter Wright-Bevans and Murray consider the operation of social representations of ageing and different ways of challenging these. Negura et al. (2018) in their discussion of the operation of social representations argue that it is through the subjective internalisation of our relationship with the social world that power operates. Power does not operate by force but by acceptance of ways of seeing the world. Individual and collective stories are at the centre of social representations (see Laszlo 2008). These narrative accounts are not only involved in the organisation of social representations but also shape our expectations and social practices (see also Murray 2002). At the core of social representations of ageing is a narrative of decline epitomised in the various stories about old people. However, while this might be the dominant narrative it can be challenged in various ways as is discussed in this final chapter.

An Introduction

This collection is constructed to introduce these approaches and to highlight the unique contribution of the various psychologies to our understanding of ageing. We have divided this book into the three sections of 'social and critical psychology', 'biological, cognitive and health psychology', and 'societal psychology'—and designedly so. There is both convergence and epistemological tension in this mapping of psychology's terrain. Whilst some may argue that disciplinary sub-division is unhelpful in the landscape of ageing research, we felt that creating distinct spaces for exploring research and practice not only enables a more inclusive understanding of psychological theories and approaches but widens the auspices of the field (see also Peel et al. 2016; Murray et al. 2017). Margaret Morganroth Gullette (2017, p. 16) wrote rather pessimistically that 'the field of age studies has a way to go to make ageing nonboring, nonfearful, nonstupid'. We hope that this collection demonstrates that within psychologi*es* of ageing this notion is a whisper from a distant past.

References

Bowling, A., & Dieppe, P. (2005). What Is Successful Ageing and Who Should Define It? *British Medical Journal, 331*(7531), 1548–1541.

Duggal, N. A., Pollock, R. D., Lazarus, N. R., et al. (2018). Major Features of Immunesenescence, Including Reduced Thymic Output, Are Ameliorated by High Levels of Physical Activity in Adulthood. *Aging Cell, 17*(2), e12750 UNSP.

Ebrahimi, Z., Wilhelmson, K., Eklund, K., Moore, C. D., & Jakobsson, A. (2013). Health Despite Frailty: Exploring Influences on Frail Older Adults' Experiences of Health. *Geriatric Nursing, 34*(4), 289–294 pmid:23669314

Feng, Z., Lugtenberg, M., Franse, C., Fang, X., Hu, S., Ch, J., et al. (2017). Risk Factors and Protective Factors Associated with Incident or Increase of Frailty Among Community-Dwelling Older Adults: A Systematic Review of Longitudinal Studies. *PlosOne, 12*, e0178383.

Giunta, B., Fernandez, F., Nikolic, W. V., Obregon, D., Rrapo, E., Town, T., & Tan, J. (2008). Inflammaging as a Prodrome to Alzheimer's Disease. *Journal of Neuroinflammation, 5*, 51. https://doi.org/10.1186/1742-2094-5-51

Gullette, M. M. (2017). *Ending Ageism, or How Not to Shoot Old People*. New Brunswick: Rutgers University Press.

Halliday, D., Stawski, R., Cerino, E., DeCarlo, C., Grewal, K., & MacDonald, S. (2018). Intraindividual Variability Across Neuropsychological Tests: Dispersion and Disengaged Lifestyle Increase Risk for Alzheimer's Disease. *Journal of Intelligence, 6*(1), 12.

Havighurst, R. J. (1963). Successful Aging. In R. H. Williams, C. Tibbits, & W. Donahue (Eds.), *Processes of Aging* (pp. 299–320). New York: Atherton Press.

Head, D., Bugg, J. M., Goate, A. M., et al. (2012). Exercise Engagement as a Moderator of the Effects of APOE Genotype on Amyloid Deposition. *Archives of Neurology, 69*(5), 636–643.

Hilborn, J. V., Strauss, E., Hultsch, D. F., & Hunter, M. A. (2009). Intraindividual Variability Across Cognitive Domains: Investigation of Dispersion Levels and Performance Profiles in Older Adults. *Journal of Clinical and Experimental Neuropsychology, 31*(4), 412–424.

Himmelweit, H. T. (1990). Societal Psychology: Implications and Scope. In H. T. Himmelweit & G. Gaskell (Eds.), *Societal Psychology* (pp. 17–45). London: Sage.

Lamb, S. (Ed.). (2017). *Successful Aging as a Contemporary Obsession: Global Perspectives*. New Brunswick: Rutgers University Press.

Lamb, S., Robbins-Ruszkowski, J., & Corwin, A. I. (2017). Introduction: Successful Aging as a Twenty-First-Century Obsession. In S. Lamb (Ed.), *Successful Aging as a Contemporary Obsession: Global Perspectives* (pp. 1–23). New Brunswick: Rutgers University Press.

Langer, E. J., & Rodin, J. (1976). The Effects of Choice and Enhanced Personal Responsibility for the Aged: A Field Experiment in an Institutional Setting. *Journal of Personality and Social Psychology, 34*(2), 191–198.

Larson, E. B., Wang, L., Bowen, J. D., et al. (2006). Exercise Is Associated with Reduced Risk for Incident Dementia Among Persons 65 Years of Age and Older. *Annals of Internal Medicine., 144*(2), 73–81.
Laszlo, J. (2008). *The Science of Stories: An Introduction to Narrative Psychology.* London: Routledge.
Lidz, C. W., Fisher, I., & Arnold, R. W. (1992). *The Erosion of Autonomy in Long-Term Care.* New York: Oxford University Press.
Lövdén, M., Li, S.-C., Shing, Y. L., & Lindenberger, U. (2007). Within-Person Trial-to-Trial Variability Precedes and Predicts Cognitive Decline in Old and Very Old Age: Longitudinal Data from the Berlin Aging Study. *Neuropsychologia, 45*(12), 2827–2838.
Mallers, M. H., Claver, M., & Lares, L. A. (2013). Perceived Control in the Lives of Older Adults: The Influence of Langer and Rodin's Work on Gerontological Theory, Policy, and Practice. *The Gerontologist, 54*(1), 67–74.
Morse, C. K. (1993). Does Variability Increase with Age? An Archival Study of Cognitive Measures. *Psychology and Aging, 8*(2), 156–164.
Murray, M. (2002). Connecting Narrative and Social Representation Theory in Health Research. *Social Science Information, 41*(4), 653–673.
Murray, M., & Crummett, A. (2018). Combating Social Exclusion Through Community Arts. In A. Walker (Ed.), *The New Dynamics of Ageing* (Vol. 1, pp. 129–145). Bristol: Policy Press.
Murray, M., & Wright-Bevans, K. (2017). Methods for Community Research, Action and Change. In J. Brooks & N. King (Eds.), *Applied Qualitative Research* (pp. 132–143). London: Palgrave.
Murray, M., Holland, C. A., & Peel, E. (2017). Psychologies of Ageing: Research, Policy and Practice. *Working with Older People, 21*(1), 1–3.
Negura, L., Plante, N., & Lévesque, M. (2018). The Role of Social Representations in the Construction of Power Relations. *Journal for the Theory of Social Behaviour, 48*.
Nelson, G., & Prilleltensky, I. (Eds.). (2010). *Community Psychology: In Pursuit of Liberation and Well-Being.* London: Palgrave.
Newman, A., Goulding, A., & Whitehead, C. (2013). How Cultural Capital, Habitus and Class Influence the Responses of Older Adults to the Field of Contemporary Visual Art. *Poetics, 41*(5), 456–480.
Nikander, P. (2000). "Old" Versus "Little Girl" a Discursive Approach to Age Categorization and Morality. *Journal of Aging Studies, 14*(4), 335–358.
Peel, E., & Harding, R. (Eds.). (2016). *Ageing and Sexualities: Interdisciplinary Perspectives.* London: Routledge.

Peel, E., Holland, C. A., & Murray, M. (2016). Exploring Psychologies of Ageing. *The Psychologist, 29*(4), 14–15.
Phillips, A. C., Carroll, D., Bums, V. E., Ring, C., Macleod, J., & Drayson, M. (2006). Bereavement and Marriage Are Associated with Antibody Response to Influenza Vaccination in the Elderly. *Brain Behavior and Immunity, 20*(3), 279–289.
Richardson, S., Karunananthan, S., & Bergman, H. (2011). I May Be Frail But I Ain't No Failure. *Canadian Geriatrics Journal, 14*(1), 24–28.
Rowe, J. W., & Kahn, R. L. (1998). *Successful Aging*. New York: Pantheon Books.
Salive, M. E. (2013). Multimorbidity in Older Adults. *Epidemiologic Reviews, 35*(1), 75–83.
Stuart-Hamilton, I. (2012). *The Psychology of Ageing: An Introduction* (5th ed.). London: Jessica Kingsley.
Vitlic, A., Khanfer, R., Lord, J. M., Carroll, D., & Phillips, A. C. (2014). Bereavement Reduces Neutrophil Oxidative Burst Only in Older Adults: Role of the HPA Axis and Immunesenescence. *Immunity and Ageing, 11*(1), 13.
Walker, A. (2014). Towards a New Science of Ageing. In A. Walker (Ed.), *The New Science of Ageing* (pp. 1–24). Bristol: Policy Press.
World Health Organisation. (2002). *Active Ageing: A Policy Framework*. Geneva: WHO.
World Health Organisation. (2007). *Global Age-Friendly Cities: A Guide*. Geneva: WHO.
World Health Organisation. (2015, September). *Ageing and Health*. Fact Sheet No. 404. http://www.who.int/mediacentre/factsheets/fs404/en/

Part I

Social and Critical Psychology

2

Ageing Identities in the Twenty-First Century: The Social and Practical Effects of Talk About Being Old

Christine Stephens and Mary Breheny

Ageing Identities

Societies construct identities for people according to their life stage with certain rights and responsibilities and expectations for particular behaviours. Importantly, different age groups are prescribed particular social identities which have more or less power in relation to one another (Calasanti 2007). Mackean and Abbot-Chapman (2012) note that age itself can be a risk as older people are more likely to suffer poorer health, age-related loss of social status, changes in personal relationships, and diminished social engagement. Critical social psychology broadly addresses how social expectations and values affect the positioning of people in these ways. The application of such critical perspectives, in

C. Stephens (✉)
School of Psychology, Massey University, Palmerston North, New Zealand
e-mail: C.V.Stephens@massey.ac.nz

M. Breheny
School of Health Sciences, Massey University,
Palmerston North, New Zealand
e-mail: M.R.Breheny@massey.ac.nz

E. Peel et al. (eds.), *Psychologies of Ageing*,
https://doi.org/10.1007/978-3-319-97034-9_2

areas of applied social psychology such as gerontology, includes a focus on understanding these social expectations for the purposes of action and change (Minkler 1996). There are many aspects of the social world today (including increasing inequalities and changes in the functioning of our social networks) that have important implications for the wellbeing of older people. A critical approach reveals the social construction of ageing and the ways in which older people are positioned as certain sorts of people in our societies on the basis of their age.

Later life does not inevitably bring a loss of social status. In some societies, such as that of the New Zealand Māori, older age comes with higher status, increased responsibility, and more power in important areas of social life (Durie 1999). In Western societies, older people are generally relegated to less powerful positions than other age groups, more likely to be seen as burdens to society, and may be treated with contempt and neglect (Butler 2009). Hudson and Gonyea (2012) have described how most recently, the representation of older people has shifted from 'dependents', a characterisation which was dominant well into the twentieth century, to 'contenders', outsiders increasingly in competition with younger people for scarce resources. These kinds of global constructions of a single group on the basis of age have been termed 'ageist'. Ageism was originally defined by Butler in 1969 as 'systematic stereotyping of and discrimination against people because they are old' (p. 243) which goes a long way today towards encompassing the discriminatory and exclusionary effects of dominant social constructions of ageing. However, although ageism has been well recognised, it is not well theorised. Coupland and Coupland (1994) have discussed the varied and sometimes inconsistent moral and political assumptions that underlie different public accounts of ageism in terms of prejudice, discrimination, disenfranchisement, and injustice. Resistance of ageism may also include other ageist assumptions such as universalism (assumptions that all old people are alike) and oppression. While there may never be a uniform set of criteria for its assessment, critical social psychology can provide tools to recognise the damaging effects of ageism in everyday talk. While ageist discrimination is 'widespread, generally accepted, and largely ignored' (Angus and Reeve 2006, p. 138), a social constructionist lens applied to our everyday practices can highlight the injustice and oppressive effects of particular forms of ageism, while pointing to areas for social and practical change.

Social Constructionism

From a social constructionist perspective, people negotiate a positive identity for themselves in everyday talk by using shared linguistic and social resources. Everyday language reproduces societal and individual relationships of power within specific social contexts (Burr 2015; Powell and Biggs 2000). Paying attention to the ways in which particular constructions arise, strengthen, and dominate our understandings of social relations, while alternative understandings are neglected, or fall out of favour, reveals these powerful relationships. For example, Wearing (1995) used Foucault's theorising on discourse to describe the dominant discourse of ageing in the 1990s as one which constructed older people as inferior on biological grounds. Wearing argued that this construction worked to isolate and disempower older people and showed how the subject position of one in biological decline was taken up by older people themselves to 'incorporate and perpetuate powerlessness' (p. 265). These discourses also shape the ways in which older people are studied by academics. Paulson and Willig (2008) identified four discourses available to construct ageing bodies in research literature on ageing, namely: a biological discourse which emphasises the vulnerability of the ageing body; social discourses which examine and contrast different ways of talking about ageing; a female beauty discourse which includes women's recognition of changes in appearance; and a feminist discourse which sees women's ageing in terms of the male gaze. These examples are not exhaustive but illustrate the importance of language in shaping our knowledge and perspectives of the world. These understandings support the importance of language in constructing and enacting ageing identities. Using particular discourses, or combinations of discourses, makes certain subject positions such as 'dependent elder' or 'old woman' available for a speaker. These subject positions are constituted by the rights and obligations to say certain kinds of things in certain situations (Harré and Gillett 1994). Consequently, an older person's talk about independence, or their own attractiveness, does not represent an objective or detached account of their life, but a particular version of experience produced in a specific context to present themselves as a certain type of person. Language is not neutral; it conveys morally laden messages that categorise how older people should be and act and therefore enable or constrain what can be said and done (Katz 2000). Accordingly, different ways of talking about older age provide different subject positions which have implications for identity and its material effects in later life.

Ways of talking about older people shapes access to valued social roles in later life. Ainsworth and Hardy (2004), for instance, used critical discourse analysis to show the effects of ageism in the workplace and how these effects are connected to the construction of age identities. In this example of research into older workers, the material effects of these discourses include influences on organisational practices such as recruitment, redundancy, and training practices and on wider social implications such as unemployment trends, labour force participation, and incomes. Discourses also influence how older people are treated in health care and support service settings. Even within a system designed to provide care, 'negative and nihilistic attitudes' (Reed and Clarke 1999, p. 208) towards older people shape health professionals' understandings of their needs and the appropriate ways of delivering care to meet those needs. Ageist assumptions work to exclude older people from participation as respected citizens and are the reason that older people may be discriminated against, excluded, and subordinated. In the following sections, we outline discourses of ageing which presently predominate and the social identities these discourses make available to older people. Following this, we provide a specific example, from our own research on older people's talk about leisure in New Zealand, to show the ways in which available identities shape everyday life and its opportunities.

Twenty-First-Century Constructions of Ageing

Dominant discourses of ageing are shaped by the historical, material, and social context of the time. In this section we describe how discourses of 'Dependence', 'Successful Ageing', 'Individualisation', 'Healthism', and 'Ageing Bodies' have become dominant in the context of demographic changes, neoliberal economic ideologies, and social policy shifts. These discourses are drawn upon to talk about ageing by young and old alike to make particular social identities available to older people.

Dependence

The late twentieth and early twenty-first centuries have seen a major shift in our constructions of older people and their position in society. One strong influence has been responses to world-wide population ageing. Population ageing means that there is both a rise in the average age of the population, as well as a growing proportion of older people within populations. By the middle of the twenty-first century, 1 in every 5 people will be aged 60 years or over (United Nations 2015). As people live longer, particularly in developed countries, the nature of being old is also changing. Since the 1940s older people in many countries have generally had greater life expectancy, rising standards of living, better health, and are more likely to have a longer retirement following working life (MacKean and Abbot-Chapman 2012). These demographic changes influence the social identities available to older people.

Although, arguably, increased longevity and improved health into older age is a positive change for society, population ageing has not been greeted with enthusiasm. Rather, increasing numbers of older people has been generally interpreted as a frightening demographic change with catastrophic implications for health and social care services. Robertson (1999) has labelled this characterisation of changing population structure as 'apocalyptic demography', in which an increasing older population is represented as destabilising the wellbeing of society as a whole. Martin, Williams and O'Neill (2009) are among those who have noted this new form of ageism, which is seen in media accounts using alarmist metaphors of a 'time bomb' or 'tidal wave' to describe older people about to engulf the population with their high health care needs.

Demographic calculations of population ageing have been expressed in terms of a 'dependency ratio' which is the ratio of those past state pension age to working age members of the population (Kleiman 1967). This approach to measuring the level of 'burden' of an ageing population draws upon a discourse of dependence, positioning all older people as non-contributing members of society who need care. These constructions are contestable. For example, Spijker and MacInnes (2013) used recent statistics to demonstrate that this is a false measure of the ageing population. Increasing numbers of older people are remaining healthy long past the nominal retirement age and contribute to family and society in many ways. Pervasive discourses of dependency suggesting alarm are damaging and have important implications for social relations between the generations and the exclusion of older people from recognition as citizens.

Successful Ageing

Changing identities for older people are influenced by governmental policy (Townsend 1986). Social policy responses to population ageing have shifted from a focus on care towards the promotion of independence, participation, and wellbeing for older people. Today, social policy in the US, Europe, and Australasia is framed in terms of promoting health and independence among older people to reduce the burden on health and welfare systems and maximise their contribution to society (Stenner et al. 2011). The 'successful' ageing model (Rowe and Kahn 1987), which focuses on older people avoiding disease and disability, maintaining high mental and physical functioning, and remaining socially engaged, has been an important influence on research, intervention, and public policy around ageing. The World Health Organisation's (WHO) Active Ageing policy framework (World Health Organisation 2002) has further influenced the shift to policies described as 'successful', 'active', and 'healthy' or 'positive' ageing strategies.

Changes from the policy focus on decline and dependence to a focus on participation and active contribution are beneficial for many older people. However, several critiques (e.g., Estes et al. 2003; Martinson and Berridge 2015; Minkler and Estes 1999; Portacolone 2011; Rubinstein and de Medeiros 2015) have shown that the successful ageing model is problematic owing to a focus on individual responsibility for wellbeing, support for oppressive ideals, and homogenising ageist discourses (Stephens 2016). A 'successful ageing' discourse positions older people as personally responsible for engaging in exercise, diet, and social engagement prescriptions to produce good health. Policies based on these ideals are oppressive because they take little account of lifelong inequalities and different opportunities for different groups of people to achieve success. The suggestion that successful ageing can be achieved by all works very well for those who are already advantaged, while those who are already disabled or already suffering health and economic disadvantage are seen as ageing unsuccessfully.

This discursive construction encourages surveillance and blaming. In a study in which 48 people aged 55–70 years (who had already answered a questionnaire about their financial circumstances) were interviewed about their family life (Breheny and Stephens 2010), we found that those who were financially secure looked forward to a future located firmly within discourses of successful ageing. They saw their comfortable and active life as a reward for hard work and contributions to society. Their accounts obscured the financial security and structural advantage that supported their successful ageing and foregrounded their individual efforts to age successfully. Poorer participants also worked hard to situate themselves in terms of a discourse of successful ageing. In the context of poor health and poverty, these descriptions focused on taking responsibility for their circumstances and being positive. For example, one participant lived in economic hardship with many illnesses including a twisted spine that she attributed to years bent over a sewing machine for long hours and an abusive husband. She saw her future as one of deteriorating health. Nevertheless, her account of ageing focussed on remaining relentlessly positive and the interviewer praised her for this attitude:

> Int: Well you've got a lovely attitude and I think that's nice.
> Anna: Thank you, well you've got to get by don't you. (Age 65)

Just like the wealthy participants, lower income participants positioned themselves as ageing successfully through references to hard work, making good choices, and taking responsibility for outcomes. For wealthy participants, a successful ageing discourse worked to support their expectation of ongoing wellbeing and participation in later life and allowed them to position themselves as deserving, by constructing these outcomes as due to hard work. Older people who meet the ideals of successful ageing tend to take personal credit for their achievements of financial and physical wellbeing. Those who don't meet the ideals are blamed and blame themselves for their infirmity or poverty. In this way, a successful ageing discourse works to blame those in poverty and encourages independence in those who could well seek help in a kinder society.

Individualisation

Changes in population structure have occurred alongside a shift in Western countries towards differentiation of the individual life course. Expectations of appropriate life course trajectories are now less circumscribed by family and gender roles and more closely aligned with the pursuit of individually directed goals (Blaikie 1999). What it means to be a person of any age is increasingly based on living 'my life' as an individual project so that identity formation has become more intimately shaped by individual choices (Beck and Beck-Gernsheim 2002). People are expected to take responsibility for their financial, social, and physical wellbeing and manage their own lives according to individual life goals. Such expectations influence social identities. From an individualist perspective, older age reflects the culmination of a lifetime of individual choices, and older people are viewed as responsible for managing their own life trajectories.

Individualisation of biography dovetails with the dominant neoliberal economic ideology of the twenty-first century (Bell and Green 2016). The implications of these market fundamentalist ideals are more than economic. They have had a profound impact on how we understand the role of the individual in society by valorising independence, rationality, hard work, and individual responsibility for outcomes (Coburn 2000). Social policy reflects and reinforces this version of right conduct to suggest that self-care, responsible lifestyles, and constant activity are the route to individual and societal wellbeing (Katz and Marshall 2003; Pond et al. 2010). These discourses position older people who have poorer health or inadequate incomes as suffering from their own choices and lack of virtuous conduct, rather than as a reflection of social and structural inequities. Responsibility for outcomes in later life is, therefore, shifted from the collective or state to the individual.

Intergenerational Inequity

The ideological swing towards individual responsibility coupled with fears regarding the burden of population ageing has corresponded with a rise in the discursive construction of an intergenerational divide (Binstock 2010). This shift has repositioned older people as no longer 'deserving' but as increasingly burdensome consumers of health and social care services (Hurley et al. 2017). These issues are referred to by economists in terms of 'horizontal equity' and 'generational fairness' when describing the different advantages enjoyed by earlier cohorts, or the future economic burden accruing to younger generations (e.g., Bradshaw and Holmes 2013). The post-war 'baby boomer' cohort are seen as influencing social policy to their advantage at each stage of their lives, having enjoyed steadily improving living standards across their life time. In contrast, current declining home ownership for younger generations and difficulties accessing social welfare and health care are blamed on earlier 'greedy' generations (Hurley et al. 2017).

These greedy or undeserving identities are frequently described in print and online popular media today. The 'undeserving old' was identified by Fealy, McNamara, Treacy, and Lyons (2012) as one of the identities constructed in media articles about welfare provision in Ireland. The general effect of the discourses drawn on was to position older people as 'dependent' and 'other' and outside Irish society. Hurley et al. (2017) showed how such media constructions in New Zealand contribute to public debate which further polarises groups in society. Newspaper articles that discussed the future cost of pensions, and the 233 public responses to these articles, focussed on claims of intergenerational inequity which generated antagonism between respondents and widened the divisions between generational groups. In the context of rising inequalities, arguments of intergenerational inequity located older people as the cause of poor life chances for younger generations as exemplified by this respondent's post: 'Why should we have to support an early retirement age, for a generation that looted the country?' (Hurley et al., p. 12) Such generational accounting for inequality fuelled further antagonism. Older people who had been negatively positioned in terms of generational cohort sought to redress the generational balance by characterising young people as impatient and demanding: 'House prices are where they are because you oiks want everything handed to them NOW and can't be bothered saving or waiting or even in many cases working for what they need' (Hurley et al., p. 13). These divisive constructions homogenise members of each generation into single group identities with labels such as 'baby boomers' or Gen Y. Such divisions assign identities according to age alone, ignore individual differences in life circumstances, obscure intergenerational family relationships, and support shifts towards individual responsibility for provision in older age rather than collective responsibility for those who require care.

Medicalisation and Healthism

In 1989 Estes and Binney described the ways in which healthy ageing has been captured by a biomedical model which constructs ageing as a medical problem. In terms of ageing, the focus of medicalised interest is on outcomes measured in terms of disease, life expectancy, falls, multimorbidity, disability, hospitalisation, and institutionalisation. Betty Friedan (1994), an American writer, feminist and activist, described her experience of attending a medical seminar on healthy ageing in terms of her surprise that: 'they only wanted to talk about Alzheimer's, senility, and nursing homes. … Clearly they did not want to think about people over sixty-five except as helpless patients, clients of their compassionate care' (p. xxv–xxvi). A biomedical focus on prolonging healthy life is unable to encompass the acceptance of death as inevitable, and the effect of this sort of denial is to pathologise ageing bodies and position older people as 'other'. Ageing is not seen as a normal part of life, but something that can be prevented, controlled, and cured (Hodgetts et al. 2003).

The biomedical version of health promotion sees health as the product of certain behaviours, rather than as a way of being generally well. This health promotion perspective on ageing emphasises individual life style practices to prevent decline and disability. Paying attention to diet, exercise, cognitive activities, and appropriate social interaction is encouraged as part of a focus on preventing ill health and delaying death (see Chapter 6, this volume). This has been labelled 'healthism'. First described by Crawford (1980), healthism describes an individualised version of the meaning of health which encourages individuals to take responsibility for that health. Petersen and Lupton (1996) have drawn upon Foucauldian theory to describe the pursuit of health, by the active, responsible citizen who engages, by choice, in a regime of constant self-evaluation and mastery of conduct, diet, and lifestyle.

There are two main concerns about the effects of this focus on health as the product of individual behaviours. First, a moral imperative pervades: those who are healthy and diligently follow dominant health messages are seen to be living virtuously, whilst those who are unhealthy or indifferent to the prescriptions for health-related behaviour are irresponsible and blameworthy for any illnesses that develop (Crawford 2006; Lupton 1995). Accordingly, older people's talk about their health includes a moral dimension in which people position themselves, and others, as virtuous or irresponsible depending on their body's condition and how well they engage in health-related practices. Pond et al. (2010) found that older people spoke about their virtuous practices such as purchasing fitness equipment, taking up regimented exercise, and daily attention to diet; however, they also mentioned their failures, such as eating too many treat foods. Many of these participants positioned themselves as failing in some way and said that they *should* be doing more. They blamed themselves for their illnesses. People also presented themselves as virtuous by comparing themselves to others who weren't. A woman positioned herself as a good person by pointing to her older friends' bad dietary habits: how they do 'nothing but sit and eat cake', which she said 'horrifies' her (Pond et al., p. 739). Work by authors such a Crawford and Lupton has demonstrated that one of the effects of healthism is to make people feel ashamed of their own ill health. It is particularly difficult for older people to be subjected to this discipline because they are more likely to suffer a disability or general loss of physical abilities as they age.

The second problem created by a focus on physical functioning is there is no space within the discourse to include bodily ageing and the inevitability of death. A healthism discourse does not include the realities of embodied ageing, but rather denies death and constructs a version of ageing in which people may never grow old (Katz and Marshall 2003; Powell and Biggs 2000, 2004). The contradiction of these ideals and the actual reality of physical decline were beginning to be noticed by some of the participants in the Pond et al. (2010) study. For example, Douglas described working hard at good health through annual check-ups, healthy eating, growing fresh produce, and keeping fit working on his lifestyle property. Despite these efforts, he had recently noticed physical changes. In spite of this, days spent labouring around his property planting trees, digging, and chopping wood, Douglas had noted a reduction in strength as he aged. Douglas expressed a sense of dissonance between health promotion discourse and his own experience by saying: 'I've often wondered whether I should start some kind of fitness regime along those lines but it's rather annoying when you've been out working all day and you come in physically tired and you still feel that you might need more exercise. It just doesn't quite add up somehow' (Pond et al. 2010, p. 741). Douglas was surprised that despite being active he was physically changing. Subjectively, he was uncomfortable with 'accepting' the ageing process because of anxiety about going 'downhill' but also worried because the health promotion advice would be to *increase* his physical activity. There is no mention of rest and relaxation in health promotion. Suggestions that relentless activity and virtuous diets might allow us to work on and live forever create difficulties for older people who are experiencing changes in energy and strength levels.

Ageing Bodies

Older people's sense of identity must be negotiated in the context of their experiences of changing bodies and the dominant discourses described earlier (see Chapter 3, this volume). The biomedical model has provided a construction of the ageing body as inevitably declining, a construction which has become the dominant narrative of the nineteenth and twentieth centuries. Thus, the 'diverse human experience of ageing was gradually reduced to a single biomedical model of decline as defined by … medical experts' (Hepworth 2003, p. 90). Hepworth stresses that sociologists do not deny the biological changes of ageing but rather provide critical understandings of the ways in which these changes are constructed as 'decline' and the effects of these constructions. An important implication is the separation of self from the body as people struggle to maintain a coherent identity as a valued citizen. Tulle (2008) describes how bodily ageing has been theorised as a threat to self and social identity and deprived of cultural capital in a cultural context which values youth, beauty, and physical competence. The proliferation of images of idealised bodies in contemporary culture devalues the aged body. At the same time, other qualities of age such as experience and maturity are no longer respected. Old and young are placed in opposition to one another and youth is accorded primacy, and therefore 'bodily ageing operates a break with our youthful past, and this is presented as detrimental to our sense of self' (Tulle 2008, p. 4).

These discursive threats to the identity of older people fuel an anti-ageing industry which promises youthful appearance in a variety of ways. For instance, Toni Calasanti (2007) has studied online advertisements for anti-ageing products to show the aversion to ageing that they reveal. She describes two central messages: ageing is aversive and controllable and those who do not control their ageing deserve shame; what constitutes ageing is different for men and women. Calasanti shows that appearance accounts for a larger proportion of women's social capital, and much of the anti-ageing discourse includes a recognition that women's importance depends on their allure to men. Sexual invisibility is a further challenge for a woman's identity, while for men, judgements about ageing are related to performance and strength. Men risk being marginalised if they cannot display these qualities. In these ways ageism and sexism are intertwined to threaten exclusion for older people from Western societies.

The association of youthful bodies with sexuality and sexual expression is the source of an additional aspect of marginalisation for older people. Representations of sexuality as the domain of the young and those with perfect bodies in the media (Hinchliff and Gott 2016) and in practice settings such as health care (Ward et al. 2005) have resulted in older people generally being identified as asexual in Western societies, despite evidence to the contrary (Hinchliff and Gott 2016; Peel and Ellis forthcoming). The female body in particular is portrayed as particularly undesirable and older women's sexuality not taken seriously.

The effects of the discursive marginalisation of ageing bodies threaten older people's own sense of identity. Furthermore, the construction of older bodies as asexual and non-productive leads to problems of the treatment of older people in many areas of practice which are highlighted in the field of institutional care. Ward et al. (2005) describe the lack of recognition of sexuality or provision of opportunity for expression of sexuality for older people living in care. People with dementia are particularly regulated in terms of sexual expression. Furthermore, although there is growing research around the experience of lesbian, gay, bisexual, and transgender (LGBT) older people, and recognition of the issues for non-heterosexuals as they age, the discourses of elder asexuality mean that 'neither in policy nor practice does the older lesbian or gay man exist as a category of client' (Ward et al. 2005; p. 51). In these ways certain important aspects of older people's self are denied and made invisible.

Discourses Speak to Each Other

As Ian Parker (1992) has described, discourses refer to other discourses. In constructing ageing and providing identities for older people, the currently dominant discourses work together to support ideologies such as individualisation and healthism, while apparently creating a coherent ageing subject. This older person is also the object of the practices of institutions such as academia, medicine, or government. Accordingly, a medical discourse of ageing, a social policy discourse of successful ageing, and discourses of individual responsibility work well together to construct the citizen who is responsible for their own physical and financial wellbeing and who, with appropriate effort, can defy ageing itself. These discourses are supported by and reinforce constructions of ageing bodies as pathological and older people as greedy graspers of scarce resources for which others are competing. At the same time, we recognise that discourses are historically located (Parker 1992). In the twenty-first century the changes wrought by population ageing and global economic pressures bring certain sets of discourses to prominence, and these influence the social identities available to older people to negotiate their place in the world (Biggs 2005). Together, population ageing, individualisation, and healthism have altered later life from a time of withdrawal and dependency to a time for pursuing health. These changes make available a new set of identities for older people based around imperatives to live a long and healthy life, actively manage their contribution and participation, and resist decline and dependency. Meanwhile, historically dominant discourses, such as those which construct older people as deserving of support, care, and respect, remain in circulation and contribute to, or contradict, current constructions of ageing. Critical practices, such as discourse analysis, may be used to disrupt seamless constructions, note contradictions and anomalies in everyday talk, and importantly, to reveal its effects and practical implications.

Discourses Are Promulgated Through the Media

Representations of ageing are often developed in science, theory, research, and government policy and reproduced through popular cultural representations of what it means to be an older person (Laliberte Rudman 2015; Wearing 1995). Townsend's (1986) influential work pointed theorists and researchers to the key role of governmental social policy in constructing versions of old age, while the medical and psychological theories behind the successful ageing model (Rowe and Kahn 1998) have been powerfully influential on research, practice, and social policy around ageing.

The media are a prominent route by which discursive constructions are made available for consumption and debate. Rozanova (2010) found that discourses of successful ageing in the print media reproduced recognisable neoliberal policy imperatives of older people as individually responsible for ageing successfully. She identified three main themes constituted the portrayal of successful ageing in *a Canadian newspaper between* 2004 and 2006. These themes were labelled: 'Successful aging is a personal choice', 'Individual responsibility for unsuccessful aging', and 'Aging successfully by staying engaged' (p. 217). Rozanova noted that there was considerable overlap of the themes in the articles, while 'Successful aging is a personal choice' was the most common theme. All the articles she analysed suggested that a person is able to choose how they age. Furthermore, many of the articles suggested that people must take responsibility for not ageing successfully.

Similarly, recent popular media headlines encourage older people to defy old age or refuse to age, suggesting that ageing is a choice that people make. Pronouncements such as 'It's no accident that I've reached age 88 in great health', which was used to introduce an article about prolonging life (Vernon 2015), promote healthy ageing as the application of careful self-management. This online newspaper article goes on to say: 'Zinke notes that he has average genes but has just taken good care of himself. He asserts that you'll increase the odds of a prosperous retirement – and a great rest of life – if you adhere to his seven pillars of aging successfully' (p. 1). Physical changes in later life are ignored by headlines such as, 'woman swims an extra length on every birthday' (Schroeter 2017) which suggest that older people can increase their physical fitness every year. Such media representations frame ageing successfully as resisting ageing itself.

At the same time as the media valorises independence, health, and contribution, the alternative equally widespread discourses of later life as a time of dependency and vulnerability are drawn upon. Fealy et al. (2012) conducted a critical discourse analysis to reveal five distinct identity types that were available in media texts: constructing older people either as 'victims'; 'frail, infirm and vulnerable'; 'radicalised citizens'; 'deserving old'; and 'undeserving old'. These discourses of dependence, decline, and intergenerational inequity work alongside those of independence and successful ageing to shape media representation of later life (Rozanova 2010). Media representations alternate between discursive possibilities, depicting both active and independent older people as flourishing and later life as a time of physical decline into dependency (Hodgetts et al. 2003). Older people must negotiate a positive identity within this discursive landscape.

Accepting and Resisting Ageing Identities

The broad effects of these discursive constructions are the alienation of older people as 'other'; the exclusion of those who cannot achieve the ideals of active, independent, and contributory ageing; and a general ignoring of death as the end point of ageing (Stephens 2016). The discourses of ageing homogenise older people as if they are one group and impose the ideals of the advantaged upon those who have suffered a life time of exclusion and oppression. These ageist discourses obscure more fundamental issues of variations in living standards and differences in social class, ethnicity, and gender across the life course. Groups of people who are most vulnerable to inequalities are further disadvantaged by expectations to display individual, successful, and healthy achievements. Older people take up the positions provided by the discursive constructions, or resist them in the light of their own situation and embodied experiences.

Certain constructions of ageing may dominate the discursive landscape, but, people can and do resist the identities provided. For example, in a study of constructions of dementia, Peel (2014) found that media accounts drew on two main discourses of a catastrophic epidemic and individual responsibility for prevention of dementia. A caregiver interviewed in this study resisted the healthism discourse by saying: 'there's an implication there that it's somehow your fault if you get dementia' (p. 896). Other studies find both acceptance and resistance of dominant discourses. A good example is provided by Lin et al. (2004) who examined the age identities of people engaged in an online discussion forum for older adults. Participants in the online discussions drew on both positive and negative discursive constructions of ageing to position themselves and to resist being positioned by others. Some positioned themselves as physically declining and subsequently losing engagement with society: 'Just how can one grow old gracefully and be happy when one has so little and no one to talk to?' (p. 269). Tulle (2008) noted that this approach accords with the dominant discourse of old age which deprives the body of its social capital and restricts agency to retirement and doomed attempts to defeat inevitable processes of bodily change.

Others in the Lin et al. (2004) study constructed old age as a time of freedom to explore and experiment with personal projects, while justifying this freedom as having been earned by a lifetime of responsible behaviour: 'This is the time to defy danger' (p. 268). Members of a smaller group resisted ageing itself by simply positioning themselves as belonging to a younger age group; these speakers achieved a positive identity by denying their biological age and identifying with teenagers, young adults, or the middle-aged: 'a 24-year-old woman stuck in a 63 year old body' (p. 269). This has been characterised by Tulle (2008) as the 'mask of ageing' approach used by older people to respond to the threat to their identity and social worth posed by a visibly ageing body. Other people recognised their old age but resisted being positioned as 'old people' who are declining and a burden, by emphasising their relative good health, their mental powers, or their contributions to their community despite apparently ageing bodies. These ageing identities provide advantages for some and disadvantages for others and negotiating available identities is strategic and situated. Successful ageing discourses have promoted opportunities for older people to remain involved in their communities, to extend paid employment, and increased expectations for a productive and healthy life beyond traditional retirement age. These identities shape how people approach their everyday life in older age, a time in which many people have retired from paid work and are seen as having increased leisure.

Ageing identities have been shaped by expectations for activity, health, participation, and contribution. Leisure time and time use in retirement becomes a significant site in which older people can demonstrate the achievement of these expectations and negotiate a positive identity. Research shows that the word 'leisure' itself has been appropriated by discourses of productivity, so that older people must either demonstrate active contribution or find new ways to construct resistant identities (e.g., Burden 1999). Ekerdt and Koss (2016) enquired about retirees' use of time to study identity formation among older people whose social roles have changed; older people in this study saw retirement as a space for time management and used their talk as an opportunity to demonstrate that they were busy active people who were ageing well. They talked about both productive time and relaxation time. To examine in more detail how different identities are constructed in regard to leisure time in older age, we analysed the ways that older people talk about spending time. Interviews with 153 New Zealanders aged from 63 to 93 years provided data for this discourse analysis. Details of the method can be found in Breheny and Stephens (2017).

Research Example: A Study of Discourses of Leisure

For older people, the boundaries of leisure may be blurred by ceasing or significantly reducing paid work. For those whose obligations for work or care continue, time may be apportioned differently as patterns of time demands change. To understand how leisure is constructed in later life in different situations, we examined the ways older people talk about spending time that is self-directed. We asked: how do older people today talk about the time available apart from the constraints of work, care, and other social obligations?

We found that two central, and potentially contradictory, discourses were used to construct the use of time in these interview accounts. The first we labelled the 'productive time' discourse which was used to construct the older person as engaged in productive activities, focussed on contributing to personal and societal wellbeing, and on extending life. The second, labelled the 'personal time' discourse, provided an identity for one who was focussed on enjoying time for their own satisfaction and pleasure and conscious of the limited time left.

Productive Time Discourse

When asked how they spent their time, older people often described a round of regular activities that encompassed paid and volunteer work, social activities, travel, hobbies, and sports activities, as well as routines of self-care and care of others. These accounts resemble the lists of retirees' leisure activities in academic, policy, and media accounts of what later life should hold for older people. They construct the time available to older people as a commodity to be productively used which positions older people as potentially useful to society, rather than as a burden on the health and social care system. Using the productive time discourse, many participants provided lists of activities to demonstrate that they were busy, active citizens, keeping physically fit and socially engaged, while resisting an identity as an older person who is disengaged from community life and likely to be a burden on the health system.

The 'productive time' discourse was used to construct leisure time activity as useful, rather than pleasurable; participating in activities was often described in terms of making an effort. Both the maintenance of bodily health and the achievement of a valued social identity as an ageing citizen through community engagement and volunteering were described in terms of productive use of leisure time. Martha (67 years) exemplified this use of the discourse by recounting her list of regular leisure activities that included both physical activity and socially focused volunteer work:

> Monday mornings and Thursday mornings I do Tai Chi, two or three afternoons or mornings a week we go singing, a group I belong to go singing at old people's homes. Tuesday, Thursday nights and Sunday afternoons into the evening we're rehearsing our play. Saturday mornings I go to Tai Chi in a different place, and much earlier. Half past 7 in the morning. Oh, I don't care much for that. I'm not a lover of early rising (…) Oh it's good for my health, yes definitely.

These activities were described not in terms of enjoyment but in terms of the ways that they shape specific bodily capacities. Using this discourse, older people described the careful management of their body through determined exercise and practice and a range of leisure activities undertaken to promote health and wellbeing such as yoga, tai chi, swimming, or walking for exercise. The productive time discourse also includes social participation and social contribution through volunteering activities as an appropriate use of time. Many participants were keen to demonstrate their active, social, and contributory use of time. At the end of describing her activities one woman concluded: 'And what else do I belong to? That must be enough isn't it?' (Beatrice, 78 years). Her final comment sought the approval of the interviewer for a schedule of community activities that positioned her as a citizen productively using her time, or as Molly (aged 80) put it, having a 'purpose to your life'.

The productive time discourse provides an identity for a proudly active citizen but also includes the subject position of 'vegetable' for those who fail to work on the body adequately or fail to engage socially. People often described their fear of 'vegetating' if they did not keep active: 'I don't want to become a vegetable' said Spencer (69 years). They also condemned other older people who failed to make the effort as 'couch potatoes' or judged those who did not use their time productively as 'they're vegetating' (Tim, 69 years). This is illustrated by Ling (66 years) who contrasted her own efforts with her husband's refusal to engage in such activities:

> Trying to do a little bit of exercise, go swimming, walking. Walking round the block. (…) I've been trying to get my husband to do it but no, he's a couch potato. He likes watching his Chinese movies.

The spectre of becoming a vegetable was often used to describe the motivating force for working on social engagement and the physical body when tiredness or personal pleasures might instead prevail: 'Well I certainly don't want to turn into a vegetable' (Martha, 67 years), even though this sort of activity may not be a pleasure as evidenced by Walter (91 years) who described his gymnasium visits to maintain health in terms of 'enduring, not enjoying'.

Older people who can demonstrate that they use leisure time productively achieve an identity as an active and self-managing person who is still in the process of being and becoming, not declining and vegetating. The vegetable metaphor provides a very negative identity for those who do not keep active and creates difficulties for those who are experiencing disabled bodies and difficulties walking. Such people described admonitions to exercise more as 'all very well' for others with greater mobility. For them, walking sticks or other supports would advertise their disabilities and draw attention to their apparent failure to keep active.

This discourse also has implications for the construction of time itself. The focus of a productive time discourse is on using time, and by keeping healthy, extending time: the healthy ageing citizen invests time to produce a return of more time. Time spent with the purpose of actively contributing and maintaining health and wellbeing omits the decline of older age and death itself. Like the health promotion discourse which promises longer life for those who maintain their health, the 'productive time' discourse does not include the end of life.

Personal Time Discourse

Many participants also constructed time use in later life as personal time; activities were described in terms of the enjoyment and pleasure they bring. In response to questions regarding what older people need as they age or what is important in older age, several participants focused on passing the time pleasantly: 'The things that are important to me right now is playing, and making myself happy, as I pursue my own happiness' (Aisea, 70 years).

The 'personal time' discourse provided a focus on personal choice and freedom, and this was seen as justified by a lifetime of contribution. A key component in the personal time discourse was 'now it's time for me'. This rhetorical construction suggests that previous roles or life experiences had prevented the pursuit of pleasure. It allowed speakers to describe activities such as sports and exercise in terms of the pleasure they bring and was also used to justify abandoning any activities that were difficult or did not promote pleasure and happiness.

Furthermore, 'personal time' constructs older people as inexorably declining and highlights the need to take pleasure while it is still possible to enjoy the things one loves. Using this discourse, older people described the imperative of maximising pleasure in the context of limited time: 'Let's enjoy while there's still some time' (Vikesh, 74 years). Recognition of limited time includes both mortality and the inevitability of changes in physical health, mobility, and energy. From this perspective, this discourse was also drawn on to provide support for morally challenging pleasures like gambling or sloth, by focussing on the imperative of pleasure. For example, Bruce (77) said: 'Yesterday I went to one of my favourite places, shouldn't really go there but I do every few days, that's one of the casinos'.

This discursive construction of leisure acknowledges ageing and death, by talking about time as limited and the process of change as inevitable. Recognition of decline and mortality positions pleasure as an immediate priority in later life. In comparison with the 'productive time' discourse, rather than managing the self to produce health and increase time, this discursive construction prioritises enjoyment and includes the past rather than focussing on the future: 'I'd like to go in my garden, do my garden, you know, have time to do my garden. Sit there and reminisce what I used to do with Mum [wife] and like when we did our garden' (66 years).

The 'personal time' discourse effectively provides an identity for an individual who prioritises personal projects. An important effect of the 'personal time' discourse is the different construction of time; ageing is acknowledged and time is seen as a finite resource that is shared among valued activities.

Negotiating Identities

Although both discourses may be used by the same individual, they also have competing demands. Both discursive constructions of leisure time speak to current discourses of ageing which must be negotiated in claiming socially virtuous identities. The identity provided by the 'productive time' discourse is aligned with that promoted through the discourses of successful ageing, healthism, and contribution. This identity is used to resist being positioned as dependent. Thus, older people's accounts draw on this construction of time as a resource to achieve active ageing and construct a version of ageing in which people may never grow old. Older people with good health and physical functioning may achieve virtuous identities within this discourse, and they can position those who do not participate well as 'vegetables', so that those experiencing physical changes may avoid public displays of incapacity to avoid such censure. Similarly, Wiersma and Chesser (2011) found that leisure was a significant site of negative experiences of ageing among older men which produced a narrowing of leisure activities and opportunities as men sought to avoid the identity of an 'old man'.

The 'personal time' discourse is aligned with current constructions of the self, achieved through the pursuit of individually directed goals. It is used to resist the claims of a 'productive time' discourse by emphasising the right to personal pleasure in the face of limited time. An important difference is the inclusion of an ageing identity in the 'personal time' discourse: an inclusion that allows people to acknowledge their ageing bodies, changing physical capacity, and the imminence of death. By giving primacy to individual priorities, speakers are able to neutralise the threat of bodily ageing to self and social identity. However, from a social perspective this discourse also provides a space for challenge from the intergenerational inequity discourse, so that its users must justify their personal freedom and pursuit of pleasure and resist being adversely positioned as undeserving old people resting on the wealth held back from younger generations. The dual identity of personal pleasure seeker and deserving past contributor was also described in Lin et al.'s (2004) study of ageing identities.

Time use in later life has received considerable recent attention as policymakers seek to address the issues of population ageing. Encouraging more productive time use seems advantageous for all: the older people who gain in terms of community integration and health benefits, and their wider communities who benefit from the services older people provide and from increasing wellbeing among older people. However, our analysis points to some of the ways a discourse encouraging older people to use their time productively both constrains older people in terms of the types of activities they pursue and shuns them when they can no longer maintain the level of health and contribution required of a productively ageing citizen. A personal time discourse provides an alternative that incorporates the inevitability of decline and supports participation and pleasure while there is still time. However, we do not suggest a turn to focus uncritically on 'personal time'. Focusing on personal pleasure and choice in leisure time may not be helpful for those in poverty (Dupuis and Alzheimer 2008), those who must work, or those who have constraining caring responsibilities (Horrell et al. 2015). Older people in these situations may struggle to position themselves positively within discursive accounts of later life based on freedom and personal choice.

Conclusion

Identity in later life is shaped by discourses that locate health as an individual responsibility, and active and successful ageing as a contribution that older people can make to reduce their reliance upon younger generations. Examining the complex of discursive constructions that influence ageing identities enables a critical account of the experience of ageing.

Dominant discourses and their development and promulgation in academic research, the media, and social policy accounts shape what older people can say, how they can say it, what they can do, and what support and resources they can legitimately claim. A critical discourse approach goes beyond the analysis of talk to show how language is a crucial aspect of daily practice.

By employing these analytic tools we are able to demonstrate the ways in which language shapes research, practice, and policy. We are able to critique the theoretical models (such as successful ageing) guiding our research and show how they shape the ways in which older people are seen within society and the sorts of practices in which they must engage to demonstrate their identity as respected citizens. For instance, because of a focus on individual responsibility for health and appropriate health-related behaviours, a 91-year-old man must 'endure' the gym to demonstrate his willingness to remain healthy and not vegetate. We are also able to demonstrate the ways in which institutional practices are shaped by dominant discourses so that older people are treated as outside society. For instance, we can show how sexuality and important relationships are denied in care settings. And we are able to demonstrate the powerful ways in which social policy frameworks draw on dominant discourses to influence the provision of health care and support for whole groups of people in ways that ignore their life histories and diversity. The findings from these kinds of analyses contribute to the development of new research models, changes in institutional practices, and to new social policy frameworks which take account of the very material effects of current approaches. For example, the recently published WHO 'Framework for Policy for Healthy Ageing' (Beard et al. 2016) brings a new perspective to understanding the wellbeing of older people. This document summarises the key issues to be addressed in healthy ageing policies as the need to: consider the heterogeneity of older age; address inequities, avoid ageist stereotypes; empower older people to adapt to challenges; consider the environments an older person inhabits (rather than individual behaviour as the basis of health); and consider health from the perspective of capability to function. This new policy focus is a direct response to critical analysis of the effects of previous policies as outlined earlier. The next step is for governments around the world to shift their focus.

Thus, social psychology provides tools for developing a more nuanced and structured approach to the expression and effects of ageism in our societies, while providing avenues for change. Critical examination of the construction and effects of identities in this way has implications for the development of social policy and the practices of health professionals, gerontologists, and psychologists in the future.

References

Ainsworth, S., & Hardy, C. (2004). Critical Discourse Analysis and Identity: Why Bother? *Critical Discourse Studies, 1*(2), 225–259. https://doi.org/10.1080/1740590042000302085

Angus, J., & Reeve, P. (2006). Ageism: A Threat to “Aging Well” in the 21st Century. *Journal of Applied Gerontology, 25*(2), 137–152.

Beard, J. R., Officer, A., de Carvalho, I. A., Sadana, R., Pot, A. M., Michel, J.-P., et al. (2016). The World Report on Ageing and Health: A Policy Framework for Healthy Ageing. *The Lancet, 387*(10033), 2145–2154.

Beck, U., & Beck-Gernsheim, E. (2002). *Individualization: Institutionalized Individualism and Its Social and Political Consequences.* London: Sage.

Bell, K., & Green, J. (2016). On the Perils of Invoking Neoliberalism in Public Health Critique. *Critical Public Health, 26*, 239–243. https://doi.org/10.1080/09581596.2016.1144872

Biggs, S. (2005). Beyond Appearances: Perspectives on Identity in Later Life and Some Implications for Method. *Journals of Gerontology – Series B Psychological Sciences and Social Sciences, 60*(3), S118–S128.

Binstock, R. H. (2010). From Compassionate Ageism to Intergenerational Conflict? *Gerontologist, 50*(5), 574–585. https://doi.org/10.1093/geront/gnq056

Blaikie, A. (1999). *Ageing and Popular Culture.* Cambridge: Cambridge University Press.

Bradshaw, J., & Holmes, J. (2013). An Analysis of Equity in Redistribution to the Retired and Children Over Recent Decades in the OECD and UK. *Journal of Social Policy, 42*(1), 39–56. https://doi.org/10.1017/S0047279412000578

Breheny, M., & Stephens, C. (2010). Ageing in a Material World. *New Zealand Journal of Psychology, 39*(2), 41–48.

Breheny, M., & Stephens, C. (2017). Spending Time: The Discursive Construction of Leisure in Later Life. *Annals of Leisure Research, 20*(1), 39–54. https://doi.org/10.1080/11745398.2016.1191360

Burden, J. (1999). Leisure as Process and Change: What Do Older People Say? *Annals of Leisure Research, 2*(1), 28–43. https://doi.org/10.1080/11745398.1999.10600870

Burr, V. (2015). *Social Constructionism*. London: Routledge.

Butler, R. N. (1969). Age-ism: Another Form of Bigotry. *Gerontologist, 9*, 243–246.

Butler, R. N. (2009). Combating Ageism. *International Psychogeriatrics, 21*(2), 211. https://doi.org/10.1017/S104161020800731X

Calasanti, T. (2007). Bodacious Berry, Potency Wood and the Aging Monster: Gender and Age Relations in Anti-Aging Ads. *Social Forces, 86*(1), 335–355. https://doi.org/10.1353/sof.2007.0091

Coburn, D. (2000). Income Inequality, Social Cohesion and the Health Status of Populations: The Role of Neo-Liberalism. *Social Science & Medicine, 51*(1), 135–146. https://doi.org/10.1016/S0277-9536(99)00445-1

Coupland, J., & Coupland, N. (1994). "Old Age Doesn't Come Alone": Discursive Representations of Health-in-Aging in Geriatric Medicine. *The International Journal of Aging and Human Development, 39*(1), 81–93.

Crawford, R. (1980). Healthism and the Medicalisation of Everyday Life. *International Journal of Health Services, 10*(3), 365–388.

Crawford, R. (2006). Health as a Meaningful Social Practice. *Health, 10*(4), 401–420.

Dupuis, S. L., & Alzheimer, M. (2008). Leisure and Ageing Well. *World Leisure Journal, 50*(2), 91–107. https://doi.org/10.1080/04419057.2008.9674538

Durie, M. (1999). Kaumatautanga Reciprocity: Maori Elderly and Whanau. *New Zealand Journal of Psychology, 28*(2), 102–106.

Ekerdt, D. J., & Koss, C. (2016). The Task of Time in Retirement. *Ageing and Society, 36*(06), 1295–1311.

Estes, C. L., & Binney, E. A. (1989). The Biomedicalization of Aging: Dangers and Dilemmas. *The Gerontologist, 29*(5), 587–596.

Estes, C. L., Biggs, S., & Phillipson, C. (2003). *Social Theory, Social Policy and Ageing. A Critical Introduction*. Maidenhead: Open University Press.

Fealy, G., McNamara, M., Treacy, M. P., & Lyons, I. (2012). Constructing Ageing and Age Identities: A Case Study of Newspaper Discourses. *Ageing and Society, 32*, 85–102. https://doi.org/10.1017/S0144686X11000092

Friedan, B. (1994). *The Fountain of Age*. London: Vintage.

Harré, R., & Gillett, G. (1994). *The Discursive Mind*. London: Sage Publications.

Hepworth, M. (2003). Ageing Bodies: Aged by Culture. In J. Coupland & R. Gwyn (Eds.), *Discourse, the Body, and Identity* (pp. 89–106). Basingstoke: Palgrave Macmillan.

Hinchliff, S., & Gott, M. (2016). Ageing and Sexuality in Western Societies: Changing Perspectives on Sexual Activity, Sexual Expression and the 'Sexy' Older Body. In E. Peel & R. Harding (Eds.), *Ageing and Sexualities: Interdisciplinary Perspectives* (pp. 11–31). London: Routledge.

Hodgetts, D., Chamberlain, K., & Bassett, G. (2003). Between Television and the Audience: Negotiating Representations of Ageing. *Health* (London), *7*(4), 417–438.

Horrell, B., Stephens, C., & Breheny, M. (2015). Capability to Care: Supporting the Health of Informal Caregivers for Older People. *Health Psychology, 34*(4), 339–348.

Hudson, R. B., & Gonyea, J. G. (2012). Baby Boomers and the Shifting Political Construction of Old Age. *The Gerontologist, 52*(2), 272–282.

Hurley, K., Breheny, M., & Tuffin, K. (2017). Intergenerational Inequity Arguments and the Implications for State-Funded Financial Support of Older People. *Ageing and Society, 37*(3), 561–580. https://doi.org/10.1017/S0144686X1500135X

Katz, S. (2000). Busy Bodies: Activity, Aging, and the Management of Everyday Life. *Journal of Aging Studies, 14*(2), 135–152.

Katz, S., & Marshall, B. (2003). New Sex for Old: Lifestyle, Consumerism, and the Ethics of Aging Well. *Journal of Aging Studies, 17*(1), 3–16.

Kleiman, E. (1967). A Standardized Dependency Ratio. *Demography, 4*(2), 876–893.

Laliberte Rudman, D. (2015). Embodying Positive Aging and Neoliberal Rationality: Talking About the Aging Body Within Narratives of Retirement. *Journal of Aging Studies, 34*, 10–20.

Lin, M.-C., Hummert, M. L., & Harwood, J. (2004). Representation of Age Identities in On-Line Discourse. *Journal of Aging Studies, 18*(3), 261–274.

Lupton, D. (1995). *The Imperative of Health: Public Health and the Regulated Body*. London: Sage.

Mackean, R., & Abbot-Chapman, J. (2012). Older People's Perceived Health and Wellbeing: The Contribution of Peer-Run Community-Based Organisations. *Health Sociology Review, 21*(1), 47–57.

Martin, R., Williams, C., & O'Neill, D. (2009). Retrospective Analysis of Attitudes to Ageing in the *Economist*: Apocalyptic Demography for Opinion Formers. *BMJ, 339*, b4914.

Martinson, M., & Berridge, C. (2015). Successful Aging and Its Discontents: A Systematic Review of the Social Gerontology Literature. *The Gerontologist, 55*(1), 58–69.

Minkler, M. (1996). Critical Perspectives on Ageing: New Challenges for Gerontology. *Ageing & Society, 16*(04), 467–487. https://doi.org/10.1017/S0144686X00003639

Minkler, M., & Estes, C. L. (1999). *Critical Gerontology: Perspectives from Political and Moral Economy*. Amityville: Baywood.

Parker, I. (1992). *Discourse Dynamics*. London: Routledge.

Paulson, S., & Willig, C. (2008). Older Women and Everyday Talk About the Ageing Body. *Journal of Health Psychology, 13*(1), 106–120. https://doi.org/10.1177/1359105307084316

Peel, E., & Ellis, S. J. (forthcoming, 2018). Ageing and Chronic Illness in Language and Sexuality. In K. Hall & R. Barrett (Eds.), *The Oxford Handbook of Language and Sexuality*. Oxford: Oxford University Press.

Petersen, A., & Lupton, D. (1996). *The New Public Health: Health and Self in the Age of Risk*. London: Sage Publications.

Pond, R., Stephens, C., & Alpass, F. (2010). Virtuously Watching One's Health Older Adults' Regulation of Self in the Pursuit of Health. *Journal of Health Psychology, 15*(5), 734–743.

Portacolone, E. (2011). The Myth of Independence for Older Americans Living Alone in the Bay Area of San Francisco: A Critical Reflection. *Ageing & Society, 31*(05), 803–828. https://doi.org/10.1017/S0144686X10001169

Powell, J., & Biggs, S. (2000). Managing Old Age: The Disciplinary Web of Power, Surveillance and Normalization. *Journal of Aging and Identity, 5*(1), 3–13.

Powell, J., & Biggs, S. (2004). Ageing, Technologies of Self and Bio-Medicine: A Foucauldian Excursion. *International Journal of Sociology and Social Policy, 24*(6), 17–29.

Reed, J., & Clarke, C. L. (1999). Nursing Older People: Constructing Need and Care. *Nursing Inquiry, 6*(3), 208–215. https://doi.org/10.1046/j.1440-1800.1999.00033.x

Robertson, A. N. N. (1999). Beyond Apocalyptic Demography: Toward a Moral Economy of Interdependence. In C. L. E. M. Minkler (Ed.), *Critical Gerontology: Perspectives from Political and Moral Economy* (pp. 75–90). Amityville: Baywood.

Rowe, J. W., & Kahn, R. L. (1987). Human Aging – Usual and Successful. *Science, 237*(4811), 143–149.

Rowe, J. W., & Kahn, R. L. (1998). *Successful Aging: The MacArthur Foundation Study*. New York: Pantheon.

Rozanova, J. (2010). Discourse of Successful Aging in the Globe & Mail: Insights from Critical Gerontology. *Journal of Aging Studies, 24*(4), 213–222.

Rubinstein, R. L., & de Medeiros, K. (2015). "Successful Aging," Gerontological Theory and Neoliberalism: A Qualitative Critique. *The Gerontologist, 55*(1), 34–42. https://doi.org/10.1093/geront/gnu080

Schroeter, M. (2017). Levin Woman Swims an Extra Length on Every Birthday. *Stuff.* Retrieved from http://i.stuff.co.nz/life-style/well-good/inspire-me/92369139/levin-woman-swims-an-extra-length-on-every-birthday

Spijker, J., & MacInnes, J. (2013). Population Ageing: The Timebomb That Isn't? *BMJ, 347*, f6598. https://doi.org/10.1136/bmj.f6598

Stenner, P., McFarquhar, T., & Bowling, A. (2011). Older People and 'Active Ageing': Subjective Aspects of Ageing Actively. *Journal of Health Psychology, 16*(3), 467–477. https://doi.org/10.1177/1359105310384298

Stephens, C. (2016). From Success to Capability for Healthy Ageing: Shifting the Lens to Include All Older People. *Critical Public Health., 27*, 490. https://doi.org/10.1080/09581596.2016.1192583

Townsend, P. (1986). Ageism and Social Policy. In C. Phillipson & A. Walker (Eds.), *Ageing and Social Policy* (pp. 15–44). London: Gower.

Tulle, E. (2008). The Ageing Body and the Ontology of Ageing: Athletic Competence in Later Life. *Body & society, 14*(3), 1–19.

United Nations. (2015). *World Population Prospects: The 2015 Revision*. Available from http://esa.un.org/unpd/wpp/

Vernon, S. (2015). 7 Pillars of Successful Ageing. *CBS News*. Retrieved from http://www.cbsnews.com/news/7-pillars-to-successful-aging/

Ward, R., Vass, A. A., Aggarwal, N., Garfield, C., & Cybyk, B. (2005). A Kiss Is Still a Kiss? The Construction of Sexuality in Dementia Care. *Dementia, 4*(1), 49–72.

Wearing, B. (1995). Leisure and Resistance in an Ageing Society. *Leisure Studies, 14*(4), 263–279. https://doi.org/10.1080/02614369500390201

Wiersma, E., & Chesser, S. (2011). Masculinity, Ageing Bodies, and Leisure. *Annals of Leisure Research, 14*(2–3), 242–259.

World Health Organisation. (2002). *Active Ageing: A Policy Framework*. Retrieved from http://whqlibdoc.who.int/hq/2002/who_nmh_nph_02.8.pdf

3

Ageing Sexualities Through a Critical Lens: Bringing Social Psychology, Sociology, and Intersectionality into Dialogue

Sharron Hinchliff and Andrew King

Introduction

In this chapter we explore the contribution of critical social psychology and sociology to understandings of the sexualities of older adults. Historically, psychological research on ageing has focused on decline (e.g. cognitive) and loss (e.g. of social status). And, with a few exceptions, psychological research on sexuality has tended to exclude older adults. But new directions are being made, and a growing body of research in social psychology, influenced by key theories within sociology, has taken a critical perspective to explore the intersections of older age with gender, sexual identity, and other social categories. Such research has demonstrated the diversity of midlife and beyond and enabled the voices of older adults

S. Hinchliff (✉)
School of Nursing & Midwifery, University of Sheffield, Sheffield, UK
e-mail: s.hinchliff@sheffield.ac.uk

A. King
Department of Sociology, University of Surrey, Guildford, UK
e-mail: andrew.king@surrey.ac.uk

E. Peel et al. (eds.), *Psychologies of Ageing*,
https://doi.org/10.1007/978-3-319-97034-9_3

to be heard: voices which challenge the narrative of older age as decline and loss. We draw on our research in the area of older adults and their gendered experiences of ageing, sexual practices, and sexual identities to demonstrate the understandings that critical theorising can bring.

We argue that ageing sexualities is an area that has not been fully explored within psychology despite the growing recognition of the sexual rights of older adults (Barrett and Hinchliff 2018; Peel and Harding 2016). While psychology has lots to offer here, we believe that ageing sexualities are better understood through a cross-disciplinary lens. The format of the chapter reflects this, starting with insights from psychology (e.g. into stereotyping and ageism) to provide background before moving to a more detailed discussion of the sexual practices and sexual identities of older adults and the tensions that currently reside in these areas. Following that, the next section provides an example of how intersectional narrative analysis can be a useful approach within critical social science. We begin, however, with a description of how we understand sexuality in this chapter as this contextualises the remaining chapter.

Positioning Sexualities

In 1969 the psychiatrist Dr Reuben published the self-help book 'Everything you always wanted to know about sex, but were afraid to ask'. It was an instant hit and a number one bestseller in over 50 countries; its reach cannot be underestimated. The chapter on menopause is particularly interesting because he wrote: 'Without oestrogen, the quality of being female gradually disappears. The vagina begins to shrivel, the uterus gets smaller, the breasts atrophy, sexual desire often disappears, and the woman becomes completely desexualised' (Reuben 1970, p. 288). He went on to say that 'it is a little worse than that' because, alongside these changes, a menopausal woman grew facial hair, developed a deep voice, and thus became as close as she could 'to being a man'. But she was 'not really a man' and 'no longer a functional woman': she was destined to 'live in a world of intersex' (pp. 288–289). While it would be difficult to find extreme views like this in contemporary Western culture, the quotation provides a useful illustration of the social construction of gender, ageing, and sexuality. The point we are making here is that sexuality should always be understood in its socio-cultural location and historical period. What we understand sexuality to mean at any time or place is not fixed. Rather, it is fluid and dynamic, shaped by the dominant values and beliefs of a culture at any given time. Which, in turn, are influenced by powerful people and leading institutions.

Two of the most notable influences on the way that sexuality is understood and experienced are religion and medicine (see McLaren 1999 for a detailed analysis). In the Reuben example above, we can assume that the so-called scientifically legitimated knowledge from a medically qualified psychiatrist helped the popularity of his book. Indeed, the text on the back cover told readers that the book was based on 'the combined experiences of several thousands of patients' who 'entrusted' Dr Reuben 'with the responsibility of treating their emotional difficulties'. The combination of scientist-as-expert and people's experiences is a powerful one and is still used (and misused) today. A recent example is the case of Flibanserin: a drug alleged to treat women's lack of sexual desire. Flibanserin was problematic from the outset, in terms of its side effects, counter-indications, and efficacy. Nonetheless, attempts to secure approval from the Food and Drugs Administration in the USA were accompanied by a large-scale marketing campaign as well as the manufactured grassroots movement 'Even the Score' which called out the disparity in drug treatments for women's sexual problems compared to men's (for more information see Snyder Bulik 2015).

The intersectional approach to sexuality we take in this chapter focuses on sexuality as practice (behaviours) and categories of identity. Traditionally, older adults have been desexualised by the positioning of sexual activity, agency, and expression in public discourses as the 'province of the young'. Due to similar restrictive attitudes, older adults' diverse sexual identities have not been recognised. Older adults have been subjected to negative, and damaging, prejudices based on these assumptions. Although stereotyping can affect everyone, it comes into sharp focus for individuals who belong to a minority or marginalised group. Stereotyping exists on multiple levels, and we now turn to the psychology of stereotyping to explore this further.

The Psychology of Stereotyping

According to the Penguin Dictionary of Psychology (Reber and Reber 2001), a stereotype is a 'set of widely shared generalizations about the psychological characteristics of a group or class of people'. We tend to make assumptions that individuals belong to such groups on the basis of their outward physical appearance (e.g. skin colour and clothes) and behaviours. Generalising our beliefs about people in this way can have advantages and disadvantages. For example, stereotypes simplify the way we view our social world and help us to quickly respond to situations. However, because of our stereotypes we may ignore individual differences because we assume that everyone of a specific group shares the same characteristics, beliefs, and values. Stereotypes can shape our view of, and behaviour towards, other people but they can also influence how we view ourselves. Research has identified that we internalise stereotypes and this in turn can affect our own thoughts and behaviours (Levy 2003). However, we can resist negative stereotypes, and negotiate our way around them which is an important point that we return to (Ussher 2006; Hinchliff and Gott 2008).

One of the most challenging areas for contemporary psychology is implicit (or unconscious) bias. This form of stereotyping refers to biases which exist and operate 'without conscious awareness, intention, or control' (Levy and Banaji 2002, p. 51). Because they are unintentional, trying to control them is tricky (Blair et al. 2011). Indeed, we need to identify our implicit biases before we even start, otherwise we do not know that we have them. Implicit biases can be powerful, but by increasing awareness we can start to decrease the influence they have on the ways we connect with other people.

In the UK, ageing tends to be viewed in terms of physical and mental decline, and as above, this dominant perspective can frame an individual's attitudes and beliefs. Levy and Banaji (2002) assert that negative views about older age remain widespread in contemporary society, and one of the reasons is the way that ageing is introduced in childhood. They argue that fairy tales expose children to older characters who are 'evil and sinister' such as the Hansel and Gretel story, or 'weak and gullible' like the grandma in Little Red Riding Hood. Levy and Banaji also highlight the lack of positive role models in television, where older actors reinforce stereotypes of decline and 'incompetence'. Messages such as these, alongside those we see in everyday advertisements and television shows which mock older adults or portray them in age stereotypical ways, contribute to implicit ageism.

Ageing and Ageism

Ageism has been described as a 'serious social prejudice with significant consequences for older people' (Age Platform Europe n.d.). The human population across the world is living longer, and at the time of writing this chapter there are more people in the UK aged 60 and older than those aged 14 and younger (Government Office for Science 2016). The forecast is that the over 60 age group will increase from 14.9 million people in 2014 to 21.9 million in 2036 (Ibid.). While the UK is undergoing a demographic shift, life expectancy varies hugely across the world. For example, average life expectancy at birth can range from 50 years in Sierra Leone to 83 years in Switzerland (World Health Organization 2016a), which has an influence on the way 'old' is understood in each country. Indeed, being old is relatively new in high-and middle-income countries as life expectancy has increased over the past 100 years.

Ageism consists of attitudinal and behavioural components, is rooted in stereotyping, and is the term used to describe prejudice (attitudes) towards and/or discrimination (behaviours) against an individual based on their age. Examples include thinking that all older adults are lonely, frail, in poor health, and consequently that they are a burden to society. The way that older adults are talked about often associates them with powerlessness and dependency. And such ageist discourses intersect with other damaging discourses (e.g. sexist, homophobic) creating multiple stigma for the older person. Ageism can manifest on many levels, from not being invited to a job interview to not being given the same medical advice about a sexual health condition as a younger person. The contributions that older adults make towards society, such as their expertise and skills which have developed over years and the provision of care for grandchildren and partners, are rarely acknowledged (see Stephens and Breheny, this volume).

Ageism is widespread. In 2016, the World Health Organization (WHO) analysed data from the World Value Survey (completed by 83,000+ people from 57 countries) and identified that 60% of respondents felt that older adults were 'not respected': a finding more likely to be reported in high-income countries (WHO 2016b). Ageism is entrenched in many societies and taken-for-granted so we are not always attuned to it. Research has identified that people with ageist attitudes tend to 'overestimate, overgeneralize, and overaccommodate the extent of actual impairments and difficulties' (Richeson and Shelton 2006, p. 184). Worryingly, ageism is more socially tolerated than racism and sexism, with many arguing that it is 'socially condoned' (North and Fiske 2012, p. 982). Indeed, ageism has been found to be the 'most tolerated form of social prejudice' in Canada (Revera and IFA 2012). And in Europe, Age UK (2011)—after analysing data from the European Social Survey—identified that reports of unfair treatment based on older age were more prevalent than those based on sex or race (35%, 25%, and 17% respectively).

The effects of ageism are powerful. The WHO (2015) report on ageing and health disclosed the myriad ways in which ageism can be bad for health and well-being. It identified key issues: that the attitudes of health professionals towards older age influenced their clinical practice; that health professionals may not be trained to deal with the complexities of aged health; and that health rationing occurred based on patient age. Medical discourses tend to pathologise older age by presenting it in terms of decline, disease, and atrophy. The focus on clinical gerontology within medical school training serves to reinforce assumptions that being old is about being frail and in need of help. Perceptions of ageing share an intimate relationship with broader cultural assumptions about health and ability and the value placed on these. The process of socialisation enables us to acquire knowledge about culture and, as psychological theory asserts, we internalise to varying degrees the messages and dominant regimes of knowledge that surround us from childhood. A combination of forces thus shapes our values, beliefs, and attitudes towards older adults.

The Shift: Positive Ageing

North and Fiske (2012) argue that the increasing older population will make their voices heard which will 'inevitably change ... social structure-based perceptions of older people' (p. 7). While ageism remains, there has been a shift in social attitudes towards older adults in Western societies over the past 50 years. This has been advanced by gerontologists, academics, and older adults themselves who have been vocal about the positives of being old and the value that older adults bring to our communities. Such advocates have challenged the overwhelming negative representations of older adults that have been dominant and fought to move away from associations of ageing as decline and burden (Bayer 2005). The cross-fertilisation of ideas has been referred to as the 'positive ageing' movement.

What is meant by positive ageing can differ. The Australian Psychological Society describes positive ageing in terms of maintaining a positive attitude, keeping physically and mentally healthy, and engaging fully in life. While for others it is focused on changing public attitudes towards ageing and ensuring age-friendly societies (e.g. the Beth Johnson Foundation). Whichever way the term is understood by organisations, making sure that older adults are valued and treated with respect and dignity are central features. A large number of bodies, including activist and grassroots groups, charities, and health authorities, have a positive ageing initiative or strategy. For example, both New Zealand (Ministry of Social Development 2001) and Ireland (Department of Health Ireland 2013) have positive ageing strategies which set out their goals for age-related policy and service delivery.

Positive ageing can, by celebrating later life in all its diversity, change the way that older age is understood, experienced, and responded to. It provides us with new discourses of ageing, ones that do not reinforce the message of ageing as decline, and a framework from which to tackle ageism. However, positive ageing is closely aligned with 'successful ageing' and the terms are used interchangeably. When this occurs, problems arise because the concept of successful ageing is intimately connected with neoliberal ideologies of health. In this chapter we use the term neoliberalisation to refer to a 'mode of governmentality that operates across a range of social spheres ... constructing individuals as entrepreneurial actors who are rational, calculating and self-regulating' (Gill 2008, pp. 442–443). And thus we view neoliberalism through a critical lens. Similarly, it is important to view the term 'successful ageing' in this way. Successful ageing refers to social, psychological, and physical well-being in older age (Carver and Buchanan 2016), but it is more than simply ageing without health problems, as we demonstrate in the following section.

Critical Social Psychology

Successful ageing and neoliberalism both approach social change by positioning the individual 'as key to social action' and, by doing so, create 'a class of unsuccessful agers' (Rubinstein and de Medeiros 2015, p. 40). The neoliberal concepts of choice, responsibility, and morality are key fodder for a critical social psychology and are discussed in the remainder of this chapter. Critical social psychology, as understood here, 'challenges social institutions and practices – including the discipline of psychology – that contribute to forms of inequality and oppression' (Gough and McFadden 2001, p. 2). Psychology has traditionally been influenced by the principles of natural science and its research methods have tended to neglect important factors that influence behaviour. Critical social psychologists argue that individuals cannot be understood in isolation from their social and interpersonal contexts and that the methods of research employed to study people should reflect this. (For further reading about critical psychology, see Murray 2014, and Gough 2017.)

Critical social psychology emphasises the influence of social norms on our behaviours, attitudes, and relationships with other people. How we think and act is influenced by cultural ideals at the micro (in-group) and macro (broader) levels. In recognition that attitudes and behaviours are deeply embedded within socio-cultural contexts, a critical social psychology asserts that the language we use supports and reinforces dominant messages and ideals. We are embedded in a web of social relations, and 'systems of difference' (e.g. class, age, gender) position us in different ways, some of which are inequitable (Gough and McFadden 2001, p. 13), but all of which intersect. Critical social psychologists are concerned with social change, and are keen to tackle oppression to improve well-being and the communities within which we live.

Critical Social Psychology on Ageing and Sex

In this way, critical social psychology has opened up new ways of looking at old age and has furthered academic theorising on the topic. It has enabled stereotypes about older adults, based on behaviours considered deviant because they fall outside of normative expectations, for example, sexual activity, to be challenged. In tackling the homogeneous view of what it means to be old, it has highlighted the rich diversity of older age. Because of the way that older age has traditionally been constructed, sexual function has been viewed in terms of decline. Sexual interest and desire have traditionally been associated with young people. And the social construction of 'sexiness' as young, wrinkle-free, and able-bodied tends to exclude the older person.

One consequence of sex being viewed as youth orientated is that older adults may refrain from expressing their sexuality for fear of being judged negatively. Such beliefs have prevented older adults from seeking help for sexual difficulties (e.g. Gott and Hinchliff 2003a). While we do not passively internalise the ageist messages of society, dominant messages establish norms and their cumulative effect can influence self-perceptions and render autonomy difficult to achieve (Nicolson 1993). Research that has explored the sexual relationships of older adults has provided evidence to challenge the view that sexual activity has no value within their lives (e.g. Gott and Hinchliff 2003b; Woloski-Wruble et al. 2010; Lee et al. 2016). Such research has been part of the driving force, as well as being driven by, a shift in social attitudes towards older adults' sexual agency. Positive ageing has been influential, and since the end of the twentieth century we have seen a significant shift in social attitudes towards older adults and sex. Indeed, sexual activity in older age is not only beginning to be widely acknowledged, it is forming 'ageing well' agendas in some high-income countries. Government sources legitimate sexual activity as part of healthy ageing by emphasising the health benefits of sex to older age. According to these sources, the benefits include an extended life expectancy, better overall health, and the ability to look younger (see Hinchliff and Gott 2016 for a more detailed discussion).

Sexual activity has come to be positioned as a necessary and healthy part of ageing through a myriad of factors. Katz and Marshall (2003) highlight the influence of sexology, with its positive sex promotion messages, and Gott (2005) argues that the backlash to the asexual old age helped drive a change to the 'sexy oldie'. These authors note that the acknowledgement of the sexual agency of older adults was significant in gerontology as it marked a shift from previous expert advice that age-related sexual decline was to be expected. The medicalisation of sex played a part as the focus on 'fixing' sexual dysfunctions accompanies the pathologisation of sexual abstinence. Sexuo-pharmaceuticals are primarily targeted towards older adults, ensuring that they do not need to sexually retire. People often experience changes in sexual function through adult life, but through the 'pharmacological imagination', standards of sexual function in older age are revisited and revised (Marshall 2010).

The cultural landscape is always changing, and other shifts around what it means to be 'older' have been observed. Older adults began to be represented in TV and film as active partners in romantic relationships (Vares 2009), which signified a change from the stereotypical roles given to an older person (e.g. grandparent). It should be considered whether the 'sexualisation of culture', where increasing attention is focused on sexual behaviours and sexual appearance by the media, the fashion and music industries, and the manufacturers that operate within this framework to sell their products, was influential here. Sexualisation is usually only considered in relation to young people, particularly women and girls, but it is not fixed and indeed can extend boundaries of age. For a discussion of the 'sexualisation' of middle-aged women see Hinchliff (2014).

The shift from sexless to sexually agentic is, at first glance, positive. It acknowledges the sexual agency of older adults and enables us to tackle ageism while creating new discourses of ageing. However, we have to ask:

who is excluded from this representation of the older person? Considered within a neoliberal framework the expectation is to age healthily and independently and thus *be* the 'virtuous citizen' (see also Chapter 2, this volume). To do so requires access to physical, social, and material resources. These resources are not always available to older adults who may have health restrictions, no sexual partner, or a disability which restricts social mobilisation. Furthermore, it excludes those who do not want to be sexually active.

The concept of successful ageing is problematic for the expectations it creates as well as the people it excludes. Indeed, sexual activity in older age is presented as a choice, as something we must do if we want to age well. It becomes, as indicated earlier, a personal responsibility and a moral imperative. The current rhetoric from the 'ageing well' agenda is to resist sexual inactivity if we want to be healthy. And as sexual activity has become a marker of successful ageing, a failure to maintain sexual agency is perceived as submitting to old age. The narratives of neoliberal health are clear to see in the 'ageing well' agendas that promote sexual activity. Those who age unsuccessfully have simply not taken up the challenge: it is their fault (Rubinstein and de Mediros 2015).

But people do not die if they are not sexually active, and sexual inactivity cannot be blamed for age-related diseases. So, we need to question whether sexual activity should be central to public health promotion in older age. Marshall (2010) thinks not, and points to the 'increased self-surveillance' and 'medical attention' to late-life sexuality that this new emphasis brings, referring to it as 'virility surveillance'. Through neoliberalism, our (older) bodies are governed, 'disciplined' (Twigg and Martin 2014) and the expectation to age well through regimes of (sexual) healthiness becomes, as Katz (2001) describes, a moral imperative. In the following section we discuss the applicability of intersectionality theory to understanding, and theorising, ageing sexualities. The method of intersectional narrative analysis is presented as a working example to highlight the complex interplay between identity, agency, and social structures.

Intersectionality and a Sociology of Ageing Sexualities

Intersectionality is an approach towards questions of identity, difference and people's lived experiences that emerged in the context of Black feminist criticisms of mainstream feminism in the late 1980s and early 1990s (Crenshaw 1993; Hill-Collins 2000). Intersectionality examines the differences within and between social groups and how power, inequality, and privilege are manifested in multiple contexts.

Since the early 2000s, intersectionality has made a significant impact on sociological studies of sexualities (Taylor et al. 2010). Intersectionality has opened up new critical space in the sociology of sexualities, offering a way to reconsider and reconcile tensions that exist between feminist and other social constructionist perspectives (Fish 2008; Jackson 2006; Richardson 2007). Moreover, it has made important contributions to exploring the relationship between class and sexuality (Skeggs 1997; Taylor 2008, 2009) and more recently it has been applied to the exploration of ageing sexualities (Calasanti 2009; King 2016a; Westwood et al. forthcoming 2018). In this respect, an intersectional approach views ageing and sexuality as co-constructing one another, such that an understanding of one is incomplete without a thorough assessment of how it is lived in relation to the other—the same can, of course, be said for how ageing and sexuality intersect with other sources of social identity, inequality, and (dis)empowerment, such as gender, class, ethnicity, and ability. Indeed, one of the strengths of intersectionality is its relentless focus on difference and diversity and unlike other approaches to ageing sexualities, such as diversity theories, it enables us to recognise and understand that these are always situated and lived within specific contexts, as well as established through social structures (King 2016a). However, as others have argued (see, e.g. the discussion in Christensen and Jensen 2012), there are theoretical criticisms of intersectionality, principally that despite its attempts to address multiple strands of identity, inequality, and (dis)empowerment, these may not manifest in the same ways, in the same contexts. In short, a more nuanced understanding of how multiple strands interact, in what contexts (and not others), is needed.

Aside from theoretical issues, one of the challenges of using intersectionality in a research context is deciding which categories are relevant in particular situations and therefore what needs to be analysed to understand (in)equalities, power, and social divisions in specific contexts. McCall (2009) is renowned for proposing a tripartite model that points to the importance of focusing on different levels: the anticategorical, the intracategorical, and the intercategorical. King (2016a) has observed how these levels correspond to the analysis of intra-group and inter-group differences, for example, exploring the differences amongst older lesbian, gay, and bisexual (LGB) people, such as those relating to gender and social class, and then comparing them to older heterosexual people.

However, researchers need to focus not only on what level to examine people's lives, identities, and the inequalities that apply to them, since intersectionality is less a methodology and more a critical perspective towards methods. So, although it is useful and arguably helps us to understand ageing sexualities, it must, in our view, be linked with and grounded in a specific mode of analysis. King (2016b, 2016c; King and Cronin 2010) has outlined how the conversation analytic approach of Membership Categorisation Analysis (MCA) can be used to this effect, since it examines people's uses of categories of identity and the actions and attributes that are associated with them in specific instances of talk. Rather than imposing an interpretation of what is important, MCA enables the researcher to explore these categories and attributes in interaction, that is, how people use them themselves. Others, however, have usefully employed an intersectional narrative analysis (INA) (Krekula 2007; Prins 2006; Christensen and Jensen 2012; Ludvig 2006), which focuses on 'not only constructions of identities, but also the role that social structures play in people's lives' (Christensen and Jensen 2012, p. 114). Thus, INA has the potential to focus on individual agency, played out in specific contexts, whilst remaining attuned to wider social structures and how they impinge upon and shape those contexts and people's lives. Moreover, within INA there is a concern to demonstrate the multiplicity of narratives; so instead of demonstrating how sexuality, for instance, is at play in a specific narrative, INA shows how it intersects with others, such as age, gender, and class as they are narrated.

Research Example: An Intersectional Narrative Approach to Ageing Sexualities

How might intersectionality, coupled with INA, enable us to examine older adults' sexuality in a more critical and less one-dimensional way, one that avoids a reductionist focus on purely the biological, or indeed one that considers only individual psychology or alternatively focuses too broadly on social structures of power? To demonstrate this, in the remainder of this chapter we apply INA to data collected in several projects[1] concerning older adults' sexuality, particularly their sexuality as identity, and discuss some key themes: the slipperiness of sexual identities; challenging age and ageism; and ageing sexualities at the intersection.

The Slipperiness of Sexual Identities

One of the most significant lessons from intersectionality theory, but also other approaches within the sociological study of sexuality, is that sexual identities are not fixed, stable, and monolithic but highly situated and related to the context in which people experience the emergence of their sexuality. Previous research has suggested that older LGB people, in particular, struggled to articulate their sexualities earlier in their lives because they were socialised and came of age in a much more prejudicial era, which has affected their life course (Knauer 2011; de Vries 2014). This can be heard in this example, from an interview conducted with Graham, who was 57 years old when interviewed:

Graham: Well I would describe myself as being bisexual with a leaning towards the male, gay side.

AK: Ok, how long have you identified as bisexual?

[1] The narratives discussed in this chapter were collected as part of research conducted by Andrew King (in collaboration with Ann Cronin) into the lives and experiences of older LGBT people living and working in a large British city. More details about the sample can be found in King, A. 2016a. *Older Lesbian, Gay and Bisexual Adults: Identities, Intersections and Institutions*. London: Routledge.

Graham: Since I was about 10 years old I suppose.

AK: Ok, have you told other people about that? Is that something you were open about when you were quite young or..?

Graham: When I was young it was illegal so it was definitely not something one discussed at school or home, definitely not with my parents anyway. So I had to be secretive about it and I've only become more open about it in my later life when things have become more, what's the word I want, generally more open.

There are two important intersectional aspects to Graham's narrative. Firstly, he corrects the interviewer—whose question concerning the temporality of his sexual identity, his openness with others, does not make reference to historical context; in this respect, Graham can be heard as emphasising that his possible lack of openness was associated with legality. Indeed, he temporises this himself with reference to becoming open about his sexuality later in his life, 'when things have become … generally more open'. Secondly, Graham specifically qualifies his use of the category bisexual—extending this with information about the gendered direction of his bisexuality. Hence, in this short section of the opening of Graham's narrative, he has already indicated that his classification as a bisexual person is intersected by law and age and is not straightforwardly classifiable.

In contrast, April, who was 59 when interviewed, spoke about how she identified herself as a lesbian providing a different form of intersecting context:

> I now identify as lesbian, but cannot remember a particular point in my life when this became apparent to me. From a very young age I felt that I was somehow different (interested in different things and thinking about things differently from my peers), but didn't agonise about it. It seemed perfectly natural to me to be attracted to women and I was victim to the usual schoolgirl crushes. Having found out about one of these my mother assured me that it would be a passing phase. I did not believe that something so

intense and wonderful would be, and it wasn't. I had a physical relationship with a school friend at the age of 15, which was based on affection and was spontaneous, natural and fun.

It is apparent in April's narrative that she has come to identify as a lesbian, 'from a very young age', but unlike Graham who pointed out the legal climate at the time, April instead focuses on her mother's reaction to one of her 'schoolgirl crushes'. April illustrates a different form of 'coming out' to Graham, although like Graham her identification as a lesbian is not something that should be unconditionally assumed.

Indeed, the vast majority of the participants in the studies we discuss in this section preferred to use their own terms or qualify the use of sexual identity categories. These included 'homosexual', 'dyke', 'iron', 'queer' to 'chicks at 56'—a reference that two lesbians who lived at a particular house number used, which was a 'nickname' given to them by a male neighbour. The point worth noting here is that these categorisations/identifications were situated within their biographies, often in relation to a wider social structural context and frequently intersected by other forms of social division and in/equality. Adhering to LGBT was, for these participants, something they did when necessary; it was not a given.

Challenging Age and Ageism

As we noted earlier, older adults' sexuality is frequently subject to ageist assumptions and narratives of inevitable decline (sometimes including their own). Whilst this is problematic for all older adults, older LGBT people are rarely represented in popular culture—when they are it is often in terms of a sad, lonely, and isolated later life (King 2016a; Traies 2012). Furthermore, ageism can affect how older LGBT people are treated by health service providers, not seen as 'at-risk' for some conditions because of stereotypical understandings of their sexuality and age group (Emlet 2006; River 2011; Asencio et al. 2009). Moreover, like mainstream society, LGBT communities can be ageist, especially community spaces such as social groups and commercial venues (Jones and Pugh 2005; Simpson 2013; Traies 2016).

The narratives of participants in the studies we discuss here contained discussions of age and ageism in a variety of ways. Sometimes participants felt that their awareness of their age was only heightened in certain contexts. For example, Brian who was aged 54 years at the time of interview, talked about how he felt he was viewed by younger gay men in commercial pub venues: 'I've noticed it in the pub, as I say younger people, they will acknowledge you in a way, but not the way they acknowledge each other'. Meanwhile, Abbey expressed how she became aware of her own age only when interacting with younger lesbians: 'I think the only time I think about that is when we are surrounded by the younger lesbians'. Moreover, Ernest, who was 72 years old at the time of interview, directly challenged the interviewer's attempt to categorise him as an older gay man. After firstly rejecting his own categorisation as gay, he went on:

> Not classifying myself as gay, I don't classify myself as old. I just don't think in terms of age, although through community organisations I have become a lot more aware … that's brought me more in to focus the needs of older people and what older people talk about, which is mainly sitting around chatting about the old days (laughter). It's not really my scene but you know, you listen and you try and be as helpful as you can.

In some ways, Ernest's narrative draws on an ageist discourse itself; he uses a stereotype about older adults reminiscing. Other participants made disparaging remarks about other LGBT people they felt were not 'acting their age', that is, not behaving or dressing appropriately; in short, a form of self-imposed ageism. Hugh, who was 54 years old when interviewed, stated:

> You know we all do it, I've seen people, I went to [club] with my flatmates the other day and there were people in there who were dressed quite inappropriately for their age, they were quite possibly younger than me but they were dressing like teenagers!

However, other participants expressed a concern with suppositions made by others, particularly professionals about the intersection of their sexuality and age. This was particularly pertinent when discussing medical professionals. Participants recounted being invisibilised as older LGB people because these professionals could not envisage older adults being anything other than heterosexual. Aside from the ageism involved here, the intersection of a powerful heteronormativity was also evident, as in the following quote from Pierre's narrative where he was discussing a visit to a general practitioner when suffering from an HIV-related illness.

> I was diagnosed with pneumonia and I was like, my third week and I needed to get more antibiotics. I didn't have a GP so I thought this is the one near to me, and then he said a question: 'Why you not married you're 40 years old, why you not married and what do you do? You're a dancer, mmm'. He was an awful, awful man. I never saw him again.

These narratives clearly show that intersections between age, ageism, and sexuality are complex—participants experienced ageism from others, sometimes they used ageist discourses themselves and many positioned themselves as 'not old' in ways that suggested that to be old and lesbian, gay or bisexual was either incompatible or socially stigmatised. In all cases, we can see that there is not a clear-cut and uniform relationship between sexuality and ageing—these intersections are always situated and can be empowering and/or disempowering.

Ageing Sexualities at the Intersection

One of the strengths of intersectionality and INA is its requirement to focus on a range of intersections, as they are manifested in a participant's narrative, rather than just one or two axes of identity, such as sexuality and age. In this case, to focus on how sexuality and age are co-constructed in relation to others. This is important because ageing sexualities will be experienced differently depending on these other forms of social division, such as, but not limited to, gender, ethnicity, and social class. In this section, we provide examples to demonstrate these complex intersectional aspects of identity.

Demonstrating the significant intersections of gender and class to ageing sexualities are the following two narratives, given by Leonard, a

63-year-old gay man, and Maz, a 54-year-old lesbian, which we are quoting at length and simultaneously here to illustrate these contrasts. Both have been asked about retirement and any plans they might have:

> I'm very fortunate in that I did a job that I thoroughly enjoyed doing. It gave me lots of opportunities, it was very well paid and it also gave me lots of lifelong friends and most of my friends have remained my friends. … Fortunately I'm, as you've asked me about myself, I'm quite financially solvent so I have opportunities to do various things or as my father rather succinctly put it, I can afford to do the things that I want to do but I can't afford to do the things that don't want to do anyway so that basically sums up, so I have quite a comfortable sort of retired life. (Leonard, gay man, 63)

> I'd love to have the time to spend in the garden and get the garden, you know, under control and enjoy being in it. I would like to be able to travel because I've not been in a position to do that. I've never had any money, and its only since I've been working full-time the last few years that I've had any money at all and what I have now is, you know, sort of split … for the mortgage and about a third for my pension … which I didn't have at all because it was with my husband and so that's all gone, so I have nothing, and about a third which is … living and … a big chunk towards saving things and pension and trying to save some so that I've got something to live on, because otherwise I'm going to have nothing, and a big chunk goes towards supporting the kids. (Maz, lesbian, 54)

These narratives demonstrate that Leonard and Maz have very different economic resources at their disposal, which can be utilised in their retirements, but these differences are strongly associated with their gender and sexuality. Maz had previously been in a heterosexual marriage and came out as a lesbian later in life. Her social network was composed of a few lesbian friends, her children, and work colleagues. Leonard, in contrast, had lived most of his adult life as an openly gay man and had a wide network of gay male friends. Asked to classify their class positions—Leonard opted for middle class and Maz for working class. In Maz's case gender, class, and sexuality were disempowering her choices about retirement, whereas Leonard's were not. However, we should not assume that this gender-class-sexuality-ageing axis is the norm—in the studies we discuss here, some working-class gay and bisexual men were not as economically solvent as some of the older lesbians; yet they do have access to a more privileged gender identity.

As Suen (2015) has noted, ethnicity is frequently ignored in research on ageing sexualities whether that is because research samples are not ethnically diverse, or whether the issue of ethnicity is never discussed—a form of triple invisibility takes place leaving studies of ageing sexualities making universalist claims from ethnically homogenous samples. Indeed, it is notable that this implicit universalism was at the very centre of intersectional critiques of feminism, made by Black Feminists in the late 1980s and early 1990s (Crenshaw 1993; Hill-Collins 2000). In the LGBT studies we discuss, the topic of ethnicity emerged from participants' narratives in a number of ways, but particularly in relation to ethnic 'others'—those of different ethnic or religious-cultural groups and their perceived attitudes towards sexual minorities.

In the following quote from Sandy's narrative, she compares her experience of a local social group for older adults with that of another person who attended, remarking upon this woman's ethnicity: 'There was someone else who I felt didn't get included and that was a black woman. I think gay and black people often have a lot in common on that level' (Sandy, lesbian, 64). In this way, Sandy was making a connection, a point of commonality, between her experience as a woman marginalised because of her sexuality to another who she felt was marginalised by her ethnicity. Regardless of whether the statement is 'true', the Black woman in question may or may not have felt excluded; this illustrates what Sandy believes are intersecting factors in 'fitting in' or not, in a mainstream older person's group. So, whilst sexuality may be one factor, ethnicity can be another. What is, however, implicit here is Sandy's unremarked upon 'whiteness'—and this was something that was apparent in interviews with other participants.

Participants frequently made distinctions between themselves (and their sexuality) and the local Muslim community in terms captured here by Ernest:

> I've become much more aware of the influence of the Mosque and the Muslim community. So I think I've always had a bit of a feeling that the

> Muslim community might be more homophobic than the non-Muslim community. So I've always been a bit more cautious. I have Muslim friends and I've always been surrounded where I particularly live at the moment, by the Muslim community. I've always done my best to be as friendly and open towards them and respect them as my fellow citizens and hope they would do the same for me. (Ernest, gay man, 73)

There is an aspect of Ernest's talk that represents an implicit Orientalism (Said 1979) that posits an 'us' (gay/bisexual people) and 'them' (Muslims) as diametrically opposed, even if in this case there can be 'respectful' co-existence. A problem with this, noted by others, is that such discourses can be utilised to portray Islam as regressive and 'medieval' (Mepschen et al. 2010).

Even when these intersections were reflected upon and viewed as more complex by participants, religiosity, ethnicity, and sexuality were somehow viewed as incompatible:

> I don't think [local area] is any worse than anywhere else, but because they are within their own religious structure it's not, well it's not in Christendom, is the certain interpretations of religious documents is that it's taboo and you don't do it, but it's far more in that, in the Muslim belief. Probably for Asian men to come out and just, I should think it's actually very hard. (Leonard, gay man, 63)

Here Leonard does at least recognise that 'Asian men' may be gay or bisexual, but this is contrasted, implicitly, to a universalised Christian perspective where although 'coming out' may be difficult, it is seen as less difficult than in Islam. We are not trying to suggest that these participants are xenophobic, racist or Islamophobic. However, they do demonstrate that ageing sexualities are being co-constructed in relation to categories of ethnicity, as well as gender and class noted earlier and that to ignore this—or not pay it sufficient attention—is problematic. It is our contention that further categories of identity, often made relevant by participants themselves in their narratives, cannot be ignored and doing so risks obscuring the complexity and multidimensionality of ageing sexualities.

Conclusion

As we have seen, ageing sexualities are subject to many assumptions and consequently shrouded in misunderstanding. Research from a critical standpoint which adopts intersectionality as an approach can help to further our understanding and challenge the stereotypes which render older age sexual agency and identity invisible and taboo. Recommendations for psychologists whose work brings them into contact with marginalised groups include raising awareness of the ways that stereotyping can affect our behaviours, for example, through educational programmes which include implicit bias training. Ensuring that appropriate policies are in place to address the sexuality-related needs of older adults is a key recommendation too. Such a policy would outline the sexual rights of older adults and suggest ways for practitioners to adopt them (see Barrett and Hinchliff 2018). It is our contention that older adults themselves should be central to discussions so that user-focused agendas are established.

References

Age Platform Europe. (n.d.). *Join WHO Global Platform Against Ageism*. https://www.age-platform.eu/policy-work/news/join-who-global-campaign-against-ageism. Accessed 09 May 2017.

Age UK. (2011). *A Snapshot of Ageism in the UK and Across Europe*. Retrieved from: http://www.ageuk.org.uk/Documents/en-GB/ID10180%20Snapshot%20of%20Ageism%20in%20Europe.pdf?dtrk=true. Accessed 09 May 2017.

Asencio, M., Blank, T. O., Descartes, L., et al. (2009). The Prospect of Prostate Cancer: A Challenge for Gay Men's Sexualities as They Age. *Sexuality Research and Social Policy, 6*(4), 38–51.

Barrett, C., & Hinchliff, S. (2018). *Addressing the Sexual Rights of Older People: Theory, Policy and Practice*. London: Routledge.

Bayer, K. (2005). Cosmetic Surgery and Cosmetics: Redefining the Appearance of Age. *Generations, 29*(3), 13–18.

Blair, I. V., Steiner, J. F., & Havranek, E. P. (2011). Unconscious (Implicit) Bias and Health Disparities: Where Do We Go from Here? *The Permanente Journal, 15*(2), 71.

Calasanti, T. (2009). Theorizing Feminist Gerontology and Sexuality: An Intersectional Approach. In V. L. Bengtson, M. Silverstein, N. M. Putney, et al. (Eds.), *Handbook of Theories of Ageing* (pp. 471–486). New York: Springer.

Carver, L. F., & Buchanan, D. (2016). Successful Aging: Considering Non-Biomedical Constructs. *Clinical Interventions in Aging, 11*, 1623.

Christensen, A.-D., & Jensen, S. Q. (2012). Doing Intersectional Analysis: Methodological Implications for Qualitative Research. *NORA – Nordic Journal of Feminist and Gender Research, 20*(2), 109–125.

Crenshaw, K. (1993). Mapping the Margins: Intersectionality, Identity Politics and Violence Against Women of Color. *Stanford Law Review, 43*(6), 1241–1299.

de Vries, B. (2014). LG(BT) Persons in the Second Half of Life: The Intersectional Influences of Stigma and Cohort. *LGBT Health, 1*(1), 18–23.

Department of Health Ireland. (2013). *The National Positive Ageing Strategy*. Retrieved from: http://health.gov.ie/healthy-ireland/national-positive-ageing-strategy/. Accessed 28 June 2017.

Emlet, C. A. (2006). 'You're Awfully Old to Have This Disease': Experiences of Stigma and Ageism in Adults 50 Years and Older Living with HIV/AIDS. *Gerontologist, 46*(6), 781–790.

Fish, J. (2008). Navigating Queer Street: Researching the Intersections of Lesbian, Gay, Bisexual and Trans (LGBT) Identities in Health Research. *Sociological Research Online, 13*(1). Retrieved from: http://www.socresonline.org.uk/13/11/12.html. Accessed 21 Nov 08.

Gill, R. (2008). Culture and Subjectivity in Neoliberal and Postfeminist Times. *Subjectivity, 25*(1), 432–445.

Gott, M. (2005). *Sexuality, Sexual Health and Ageing*. Berkshire: Open University Press.

Gott, M., & Hinchliff, S. (2003a). Barriers to Seeking Treatment for Sexual Problems in Primary Care: A Qualitative Study with Older People. *Family Practice, 20*(6), 690–695.

Gott, M., & Hinchliff, S. (2003b). How Important Is Sex in Later Life? The Views of Older People. *Social Science and Medicine, 56*(8), 1617–1628.

Gough, B. (Ed.). (2017). *The Palgrave Handbook of Critical Social Psychology*. London: Palgrave Macmillan.

Gough, B., & McFadden, M. (2001). *Critical Social Psychology: An Introduction*. London: Palgrave Macmillan.

Government Office for Science. (2016). *Future of an Ageing Population*. Retrieved from: https://www.gov.uk/government/publications/future-of-an-ageing-population. Accessed 09 May 2017

Hill-Collins, P. (2000). *Black Feminist Thought: Knowledge, Consciousness and the Politics of Empowerment*. New York: Routledge.

Hinchliff, S. (2014). Sexing Up the Midlife Woman: Cultural Representations of Ageing, Femininity and the Sexy Body. In I. Whelehan & J. Gwynne (Eds.), *Ageing, Popular Culture and Contemporary Feminism* (pp. 63–77). London: Palgrave Macmillan.

Hinchliff, S., & Gott, M. (2008). Challenging Social Myths and Stereotypes of Women and Aging: Heterosexual Women Talk About Sex. *Journal of Women & Aging, 20*(1–2), 65–81.

Hinchliff, S., & Gott, M. (2016). Ageing and Sexuality in Western Societies: Changing Perspectives on Sexual Activity, Sexual Expression, and the Sexy Older Body. In E. Peel & R. Harding (Eds.), *Ageing and Sexualities: Interdisciplinary Perspectives* (pp. 11–31). London: Routledge.

Jackson, S. (2006). Interchanges: Gender, Sexuality and Heterosexuality: The Complexity (and Limits) of Heteronormativity. *Feminist Theory, 7*(1), 105–121.

Jones, J., & Pugh, S. (2005). Ageing Gay Men: Lessons from the Sociology of Embodiment. *Men and Masculinities, 7*(3), 248–260.

Katz, S. (2001). Growing Older Without Aging? Positive Aging, Anti-Ageism, and Anti-Aging. *Generations, 25*(4), 27.

Katz, S., & Marshall, B. (2003). New Sex for Old: Lifestyles, Consumerism and the Ethics of Aging Well. *Journal of Aging Studies, 17*(1), 3–16.

King, A. (2016a). *Older Lesbian, Gay and Bisexual Adults: Identities, Intersections and Institutions*. London: Routledge.

King, A. (2016b). Queer Categories: Queer(y)ing the Identification 'Older Lesbian, Gay and/or Bisexual (LGB) Adults' and Its Implications for Organizational Research, Policy and Practice. *Gender, Work and Organization, 23*(1), 7–18.

King, A. (2016c). Troubling Identities? Examining Older Lesbian, Gay and/or Bisexual People's Membership Categorisation Work and Its Significance. In E. Peel & R. Harding (Eds.), *Ageing and Sexualities: Interdisciplinary Perspectives* (pp. 163–182). London: Routledge.

King, A., & Cronin, A. (2010). Queer Methods and Queer Practices: Re-Examining the Identities of Older Lesbian, Gay, Bisexual (OLGB) Adults. In K. Browne & C. Nash (Eds.), *Queer Methods and Methodologies* (pp. 85–96). Aldershot: Ashgate.

Knauer, N. J. (2011). *Gay and Lesbian Elders: History, Law and Identity Politics in the United States*. Farnham: Ashgate.
Krekula, C. (2007). The Intersection of Age and Gender: Reworking Gender Theory and Social Gerontology. *Current Sociology, 55*(2), 155–171.
Lee, D. M., Vanhoutte, B., Nazroo, J., & Pendleton, N. (2016). Sexual Health and Positive Subjective Well-Being in Partnered Older Men and Women. *Journals of Gerontology Series B: Psychological Sciences and Social Sciences, 71*(4), 698–710.
Levy, B. R. (2003). Mind Matters: Cognitive and Physical Effects of Ageing Self-Stereotypes. *Journals of Gerontology Series B: Psychological Sciences and Social Sciences, 58B*(4), 203–211.
Levy, B. R., & Banaji, M. R. (2002). Implicit Ageism. In T. D. Nelson (Ed.), *Ageism: Stereotyping and Prejudice Against Older Persons* (pp. 49–75). Cambridge, MA: MIT Press.
Ludvig, A. (2006). Differences Between Women? Intersecting Voices in a Female Narrative. *European Journal of Women's Studies, 13*(3), 245–258.
Marshall, B. (2010). Science, Medicine and Virility Surveillance: 'Sexy Seniors' in the Pharmaceutical Imagination. *Sociology of Health & Illness, 32*(2), 211–224.
McCall, L. (2009). The Complexity of Intersectionality. In E. Grabham, D. Cooper, J. Krishnadas, et al. (Eds.), *Intersectionality and Beyond: Law, Power and the Politics of Location* (pp. 49–74). London: Routledge.
McLaren, A. (1999). *Twentieth Century Sexuality: A History*. Oxford: Blackwell Publishing.
Mepschen, P., Duyvendak, J. W., & Tonkens, E. H. (2010). Sexual Politics, Orientalism and Multicultural Citizenship in the Netherlands. *Sociology, 44*(5), 962–979.
Ministry of Social Development. (2001). *New Zealand Positive Ageing Strategy*. New Zealand Government. Retrieved from: http://www.msd.govt.nz/about-msd-and-our-work/publications-resources/planning-strategy/positive-ageing/. Accessed 09 May 2017.
Murray, M. (Ed.). (2014). *Critical Health Psychology*. London: Palgrave Macmillan.
Nicolson, P. (1993). Public Values and Private Beliefs: Why Do Women Refer Themselves for Sex Therapy? In J. M. Ussher & C. D. Baker (Eds.), *Psychological Perspectives on Sexual Problems: New Directions for Theory and Practice* (pp. 56–76). London: Routledge.
North, M. S., & Fiske, S. T. (2012). An Inconvenienced Youth? Ageism and Its Potential Intergenerational Roots. *Psychological Bulletin, 138*(5), 982–997.

Peel, E., & Harding, R. (Eds.). (2016). *Ageing and Sexualities: Interdisciplinary Perspectives* (pp. 163–182). London: Routledge.
Prins, B. (2006). Narrative Accounts of Origins: A Blind Spot in the Intersectional Approach? *European Journal of Women's Studies, 13*(3), 277–290.
Reber, A. S., & Reber, E. R. (2001). *The Penguin Dictionary of Psychology* (3rd ed.). London: Penguin Books.
Reuben, D. R. (1970). *Everything you Ever Wanted to Know About Sex, But Were Afraid to Ask*. London: W.H. Allen.
Revera & the International Federation on Ageing. (2012). *The Revera Report on Ageism*. Retrieved from: https://www.ifa-fiv.org/publication/demographics/revera-report-on-ageism/. Accessed 09 May 2017.
Richardson, D. (2007). Patterned Fluidities: (Re)Imagining the Relationship Between Gender and Sexuality. *Sociology, 41*(3), 457–474.
Richeson, J. A., & Shelton, J. N. (2006). A Social Psychological Perspective on the Stigmatization of Older Adults. In L. L. Carstensen & C. R. Hartel (Eds.), *When I'm 64: National Research Council (US) Committee on Aging Frontiers in Social Psychology, Personality, and Adult Developmental Psychology*. Washington, DC: National Academies Press.
River, L. (2011). *Appropriate Treatment: Older Lesbian, Gay and Bisexual People's Experience of General Practice*. London: Age of Diversity/Polari. Retrieved from: http://www.ageofdiversity.org.uk/sites/default/files/AppropriateTreatment.pdf. Accessed 16 July 2012.
Rubinstein, R. L., & de Medeiros, K. (2015). 'Successful Aging,' Gerontological Theory and Neoliberalism: A Qualitative Critique. *The Gerontologist, 55*(1), 34–42.
Said, E. (1979). *Orientalism*. London: Routledge and Kegan Paul.
Simpson, P. (2013). Alienation, Ambivalence, Agency: Middle-Aged Gay Men and Ageism in Manchester's Gay Village. *Sexualities, 16*(3–4), 283–299.
Skeggs, B. (1997). *Formations of Class and Gender: Becoming Respectable*. London: Sage Publications.
Snyder Bulik, B. (2015). Will Marketing Make the Difference in Approval of Sprout's 'Pink Pill' for Women? *Fierce Pharma*. Retrieved from: http://www.fiercepharma.com/regulatory/will-marketing-make-difference-approval-of-sprout-s-pink-pill-for-women. Accessed 12 Oct 2017.
Suen, Y.-T. (2015). Lesbian, Gay, Bisexual and Transgender Ageing. In J. Twigg & W. Martin (Eds.), *Routledge Handbook of Cultural Gerontology* (pp. 226–233). London: Routledge.
Taylor, Y. (2008). 'That's Not Really My Scene': Working-Class Lesbians In (and Out of) Place. *Sexualities, 11*(5), 523–546.

Taylor, Y. (2009). Complexities and Complications: Intersections of Class and Sexuality. *Journal of Lesbian Studies, 13*(2), 189–203.
Taylor, Y., Hines, S., & Casey, M. E. (2010). *Theorizing Intersectionality and Sexuality*. London: Palgrave Macmillan.
Traies, J. (2012). 'Women Like That': Older Lesbians in the UK. In R. Ward, I. Rivers, & M. Sutherland (Eds.), *Lesbian, Gay, Bisexual and Transgender Ageing: Biographical Approaches for Inclusive Care and Support* (pp. 67–82). London: Jessica Kingsley Publishers.
Traies, J. (2016). *The Lives of Older Lesbians: Sexuality, Identity & the Life Course*. London: Palgrave Macmillan.
Twigg, J., & Martin, W. (2014). The Challenge of Cultural Gerontology. *The Gerontologist, 55*(3), 353–359.
Ussher, J. M. (2006). *Managing the Monstrous Feminine: Regulating the Reproductive Body*. London: Routledge.
Vares, T. (2009). Reading the 'Sexy Oldie': Gender, Age(ing) and Embodiment. *Sexualities, 12*(4), 503–524.
Woloski-Wruble, A. C., Oliel, Y., Leefsma, M., & Hochner-Celnikier, D. (2010). Sexual Activities, Sexual and Life Satisfaction, and Successful Aging in Women. *The Journal of Sexual Medicine, 7*(7), 2401–2410.
World Health Organization. (2015). *World Report on Ageing and Health*. Geneva: World Health Organization. Retrieved from: http://www.who.int/ageing/publications/world-report-2015/en/. Accessed 29 Dec 2016.
World Health Organization. (2016a). *Global Health Observatory: Life Expectancy*. Geneva: World Health Organization. Retrieved from: http://www.who.int/gho/mortality_burden_disease/life_tables/situation_trends/en/. Accessed 29 Dec 2016.
World Health Organization. (2016b). *Discrimination and Negative Attitudes About Ageing Are Bad for Your Health*. Geneva: World Health Organization. Retrieved from: http://www.who.int/mediacentre/news/releases/2016/discrimination-ageing-youth/en/. Accessed 29 Dec 2016.

4

Empathy as a Way of Acknowledging Patients' Personhood in Palliative Care Interactions

Joseph Ford

Ageing, Dying, and Palliative Care

Ageing and dying are clearly intertwined issues, albeit ones that have not always been researched alongside each other (Gott and Ingleton 2011). There are also some key parallels and overlapping topics of debate and concern between these two areas. For example, Gott and Ingleton (2011) note that models of "successful ageing" define "'success' in terms of the ability to stave off … death," meaning that "there is no place for older people who are dying except, perhaps, as cautionary tales of 'failure.'" This contemporary concern mirrors the "impersonal high-technology" (Miller 1992, p. 227) medical environment of the period after the Second World War, when an inordinate emphasis was placed on curing patients and, conversely, little attention was paid to the palliative aspects of medicine. This state of affairs is reflected in several reports released throughout the 1950s and 1960s (Clark 2007), three of which, in particular, inspired

J. Ford (✉)
University of Exeter, Exeter, UK
e-mail: J.W.Ford@exeter.ac.uk

E. Peel et al. (eds.), *Psychologies of Ageing*,
https://doi.org/10.1007/978-3-319-97034-9_4

the founder of the palliative care movement, Cicely Saunders, to start gathering evidence and, eventually, to open St Christopher's (the "first modern hospice" (Clark 2007, p. 432)) in 1967 (Saunders 2001). As Clark (1999) notes (in language which echoes Gott and Ingleton (2011)), before the palliative care movement, there was "little attention [paid] to medicine's 'failures', to those in their last illness whose time was short" (p. 227).

Just as dying patients were neglected in the post-war era, so too is there the possibility of older, dying patients in the modern era becoming lost amidst the desire to fight off ageing on the one hand and death on the other. As Seymour and Gott (2011) note, "One of the central problems in all clinical decision-making relating to end of life care is the fluctuation and variability not only in personal wishes and preferences, but in individuals' abilities to express these." Concerns along these lines have also been raised by palliative care researchers, who have noted that dying is a highly individualised process and not amenable to one-size-fits-all conceptions. Stringer (2007), for example, shows death to be a "highly individualised concept" (p. 25). Staff and patients may differ in what they consider to be "a good death" (Payne et al. 1996), and concerns have been raised about patients feeling pressured to conform to cultural expectations about what constitutes a good death (Goldsteen et al. 2006; Hart et al. 1998; Macnamara et al. 1994). An illustration of this can be seen in Goldsteen et al.'s (2006) interview study with terminal patients and their family members. Participants acknowledged that they were culturally expected to, for example, accept or communicate openly about death. Participants did not always fit these expectations, however, with some showing a lack of acceptance or an unwillingness to communicate. As Goldsteen et al. (2006) conclude, "what can be considered a good death is something that cannot be defined beforehand … and is not the same for everyone" (p. 384).

In theory, the palliative care movement avoids these issues, built as it is around a holistic approach which takes into account not only patients' physical pain but their emotional and spiritual pain as well (Baines 2011; Saunders et al. 1995). However, there has been no prior research looking at how doctors actually put such principles into practice. Given the aforementioned possibility that older patients' personhood may be neglected, however, it is important to get a sense of how experienced palliative care practitioners actually go about treating older patients as persons with their own thoughts, feelings, and experiences. I do so here by analysing a set of interactions between doctors and older patients in palliative care, focusing particularly on how the doctors display an empathic understanding of these patients' emotions. Displaying empathy is, I show, one of the key ways in which palliative care doctors acknowledge older patients' personhood at the end of their lives.

I start by establishing my approach to empathy, which will be not as an internal mental state, but as an interactional phenomenon. I then review some of the prior work on empathy in this interactional tradition before moving on to the analysis itself. This analysis is based around a series of extracts taken from interactions between doctors and older patients in a hospice. Through these extracts, I show how doctors' displays of empathy ensure that the focus remains on older patients as whole persons.

Empathy as an Interactional Phenomenon

Empathy is a notoriously difficult concept, with several articles devoted simply to trying to define and disentangle it from the related concept of "sympathy" (see Wispé 1986). In their reviews on the subject, Hemmerdinger et al. (2007) and Pedersen (2009) show how empirical research on empathy research in medicine has overwhelmingly fallen into one of two traditions. Medical communication researchers in the first tradition have used scales and questionnaires to "measure" empathy, whether that is by asking patients to rate how empathic their doctor was in a particular interaction (e.g. the Consultation and Relational Empathy (CARE) Measure (Mercer et al. 2004)), or by asking doctors to answer a series of questions with the aim of determining how empathic they are (e.g. the Jefferson Scale of Physician Empathy (Hojat et al. 2001)).

In the other tradition, medical communication researchers have observed actual doctor-patient interactions to identify and rank empathic moments. A major body of work has, for example, been built up around the concept of "empathic opportunities" (Suchman et al. 1997), which are moments where patients say something which could, potentially, allow for an empathic response from the doctor. Researchers have developed scales to identify empathic opportunities, often with the aim of ranking the empathic quality of the doctor's response. The Empathic Communication Coding System (Bylund Makoul 2002), for example, rates doctors' responses on a six-point scale, ranging from zero (the physician's denial of the empathic opportunity) through to five (the physician shares in the empathic opportunity, offering an experience of their own analogous to the patient's).

This chapter is more closely aligned with the latter tradition than the former, inasmuch as it also uses actual doctor-patient interactions as data. However, as various critiques have pointed out (Sandvik et al. 2002; Stone et al. 2012), coding systems for empathy can be problematic because they are attempting to neatly encapsulate something that may be done in subtle ways that go beyond the boundaries of pre-established codes. Rather than identifying and ranking empathic moments, therefore, I approach empathy as something which exists within the interaction between doctor and patient. My chosen method in doing so is conversation analysis (CA), which treats interaction as an orderly phenomenon amenable to structural analysis (Sacks 1995; Psathas 1995). See Chap. 2, this volume, for discussion of other forms of discourse analysis.

The Conversation Analytic Approach to Empathy and Emotion

Conversation analysts Hepburn and Potter (2007) suggest that empathy can be approached in two ways in interaction. The first approach, similar to the one outlined earlier, would involve developing a definition of empathy and then looking for examples of that definition in interactions. The second approach (the one that they, and I, adopt), meanwhile, involves using "interaction analysis to critically explore and perhaps respecify" (p. 99) notions of empathy. In other words, rather than treating empathy as an object that *manifests* in talk, this approach sees empathy as something that is intrinsically interactional and *part of* talk.

A similar approach is outlined by Silverman and Peräkylä (1990). They note that classic accounts of empathic communication treat it as a mental process that gets communicated through talk. However, as they go on to argue, such accounts are treating public behaviours as indicators of inaccessible mental processes. The approach that they advocate is thus to set the mental processes aside and focus instead on the "logic of public actions only" (p. 313). As Ruusuvuori (2005) notes, this means seeing empathy "as [a] sequential [process] in talk-in-interaction" consisting "of turns of talk rather than a sequence of experiences" (p. 206).

The approaches advocated by Hepburn and Potter (2007), Silverman and Peräkylä (1990), and Ruusuvuori (2005) find precedent in the so-called discursive turn in psychology (see Edwards and Potter (1992) and Harré (2001)). This turn has seen classic psychological topics approached as part of naturally occurring discourse, rather than as mental states. It thus differs from traditional cognitive psychological approaches, which do indeed research internal mental processes, and use non-naturally occurring experiments to do so. It also, in its strict focus on interaction and how psychological concepts are employed within it, differs from other forms of discourse analysis (particularly critical discourse analysis) which link language to wider social structures or processes.

To give one example of such research, particularly relevant to this chapter, Edwards (1999) analyses emotion discourse, showing how emotional concepts are employed in interactions between a married couple (Connie and Jimmy) and a marriage counsellor. He shows how Connie ascribes the psychological concept of "jealousy" to Jimmy as an enduring trait and how Jimmy resists this ascription by establishing that his "anger" was a circumstantial response to provocation on Connie's part. Psychological concepts, rather than being approached simply as mental objects, are thus treated in Edwards's (1999) analysis as performing discursive actions in interaction (see also Harré and Gillet 1994).

This is in line with other "interactionist" (Hochschild 1979, 2003) work on emotion, particularly by CA researchers, which has focused less on "individuals and their emotional messages" and more on "the ways in which emotional displays are put to use and treated in the process of interaction" (Ruusvuori 2013, pp. 332–333). This has included work on how emotions are displayed in relation to other speakers' talk (Selting

2010; Wilkinson and Kitzinger 2006), research on how emotional displays (e.g. laughter and crying) can be used to perform delicate interactional tasks (Jefferson 1985), and research on how such emotional displays are responded to (Hepburn and Potter 2007, 2012; Moore 2009; Weatherall 2015; Weatherall and Stubbe 2015). (See Peräkylä and Sorjonen (2012) for a collection of CA work in this area.)

Prior Conversation Analytic Research on Empathy

So far, I have established that my approach is grounded in a discursive understanding of empathy and emotion, which treats these classic psychological topics not as internalised states but as interactional phenomena. I now review some of the empirical findings that have emerged from prior work in this tradition. Such work can, broadly, be divided into that which has examined empathy in everyday interaction and that which has researched empathy in institutional environments (including, as is the case in this chapter, medical interaction).

In the everyday interaction tradition, Gail Jefferson (1988) suggests that speakers can respond empathically when another speaker is relating a troublesome experience. This might include, for example, saying "How awful" when the other speaker reaches a particularly strong emotional point in their story. John Heritage (2011) goes further by outlining a spectrum of responses that speakers use when responding to other speakers' troubling experiences. These range from unempathic responses like asking an "ancillary question" about a side issue only tentatively related to the experience that has been described (e.g. a listener asking about a speaker's tan when the speaker has been talking about a fun holiday experience), through to highly empathic "response cries" whereby the listener simulates the response that they imagine that the speaker would issue in the situation that they have described (e.g. "Ohhh go: (d)"). Kupetz (2014), meanwhile, suggests that listeners use "response cries" and other such "fleeting" empathic responses in the early stages of a speaker's story, before moving to more substantial empathic displays as the story develops. She gives a specific example of a speaker telling her friend about the death of her grandfather. In the early stages of this story, the friend displays empathy nonverbally using facial expressions (raising her eyebrows and dropping the corners of her mouth). The friend's empathic displays become more verbal and substantial as the story goes on, however, reaching a point where, after the speaker has finished, her friend shows her understanding by talking about the death of her own grandfather.

CA research on empathy in institutional interactions, meanwhile, has been far more extensive and diverse, covering a range of environments from psychotherapy (Muntigl and Hovarth 2014; Muntigl et al. 2014; Voutilainen 2012; Weiste and Peräkylä 2014; Wynn and Wynn 2006), helpline calls (Hepburn and Potter 2007; Kitzinger 2011; Moore 2009; Pudlinski 2005), general doctor-patient interactions (Ruusuvuori 2005, 2007), HIV/AIDS counselling (Silverman and Peräkylä 1990), to disaster relief (Kuroshima and Iwata 2016) and health visitor interactions (Heritage and Lindström 2012). While a full overview of this research goes beyond the scope of this chapter, a few recurring themes from it are particularly relevant to the analysis later, and are thus worth covering.

First of all, researchers on institutional interactions have, like researchers on everyday interactions, attempted to delineate empathic conversational practices. One such practice is that of a professional "formulating" (i.e. reworking in some way) the talk of a client. Formulations are typically divided into gist and upshot formulations, with the former referring to one speaker summarising what another has said and the latter referring to drawing out the implications of what another speaker has said (Heritage and Watson 1979). The empathic potential of such formulations is that they can display an understanding of the emotional resonance of what a client has just said. While formulations are perhaps the most commonly identified way for professionals to display empathy for clients, they are by no means the only empathic practice found in the CA literature. Alternative practices include relating a similar experience (Heritage and Lindström 2012; Kuroshima and Iwata 2016; Ruusuvuori 2005; Wynn and Wynn, 2006), naming the client's feelings (Muntigl and Hovarth 2014; Hepburn and Potter 2007; Pudlinski 2005; Wynn and Wynn 2006), and assessing the client's experience (Pudlinski 2005). CA researchers have also focused on non-semantic ways for professionals to display empathy for a client's experiences, including embodied actions (such as a psychotherapist placing a hand on her client's shoulder) (Wynn and Wynn 2006), intonation (Weiste and Peräkylä 2014), and response cries (Kitzinger 2011).

CA researchers on empathy in institutional interactions have also focused on how empathy interacts with the functions that these institutions perform. Ruusuvuori (2005, 2007), for example, shows how doctors can use empathic statements in response to a patient's description of their problems to both reassure the patient and to move the interaction forward to the prescribing of a treatment. In telephone helplines, meanwhile, researchers have shown how call takers can avoid displaying empathy in direct response to callers being upset, instead leaving their empathic responses to the end of the call so as not to disrupt the flow of the institutional task (Kitzinger 2011; Weatherall and Stubbe 2015).

In this section, I outline my interactional approach to empathy. This approach involves studying empathy not as an internalised trait or process (as has traditionally been the case in empathy research in medicine) but as an interactional phenomenon. Having situated my approach within the "discursive turn" in psychology, I then outline some of the prior interactional research done on empathy, establishing a precedent for my analysis by showing how it has yielded valuable findings in the past. With all of this established, I now show how I have put this approach into practice in my analysis of doctors displaying empathy when interacting with older patients in palliative care.

Research Example: Ways of Doing Empathy with Older Patients

In the previous section, we saw multiple different definitions and conceptualisations of empathy. For the purposes of this analysis, empathy is defined as *the doctor's expressed understanding of the patient's emotional experience*. This definition was, in line with Hepburn and Potter's (2007) suggested approach, developed in tandem with data analysis rather than prior to it.

This analysis is structured around five extracts from palliative care consultations.[1] These extracts are all drawn from wider collections of the phenomena that they illustrate. There were two key criteria when selecting extracts for this chapter. The first was age—where possible, given the choice between examples, I tried to select the example featuring the older patient. It was also important, though, that the extracts be strong examples of the phenomena that they were selected to illustrate, and could also fit reasonably within the confines of this chapter (many examples would require extensive contextualisation). The selection process was, as such, a balance between these two factors.

In the first two extracts, I show how palliative care creates a space for patients to tell stories that, strictly speaking, go beyond the biomedical remit. I also demonstrate how doctors can empathise with the emotions touched off by such stories. The next extract, meanwhile, highlights how doctors can empathise with patients in the course of discussing their biomedical symptoms. The final two extracts then demonstrate how doctors can display empathy with patients' emotions when those emotions clash with the care that is being offered. The focus throughout is on how doctors' empathic displays are used to show an understanding of older patients as persons, rather than objects to be treated. This idea is developed in detail throughout the analysis.

Patients Telling Stories

Extract 4.1 comes from a consultation in which the (female) patient's friend—labelled "Com" for "Companion"—is telling the (female) doctor a story about how she found the patient struggling to breathe (the patient herself is aged 64). This was the first occasion on which

[1] The data for this chapter are taken from 37 recordings of doctor-patient hospice consultations, which were collected as part of the "Video-based communication research and training in decision-making, empathy, and pain management in supportive and palliative care" (VERDIS) Project (see https://tinyurl.com/ybk8t3mz). Researchers on this project obtained full consent from participants for the use of their recordings in research as long as the recordings were appropriately anonymised. With this in mind, pseudonyms are used in place of the patients' real names in this chapter. It should also be noted that, while the overwhelming majority of patients in the data were indeed older, this was not true in every case.

they became aware of the patient's illness, and this led to the patient's initial hospitalisation.

Extract 4.1 Oh gosh [P43/F: 9.23 - 9.53][2]

1	Com:	She just managed to eat a tiny little sandwich.=And then she
2		said she wanted to go toilet. .hhh So I says, 'Okay.' I've
3		got a chairlift anyway.=She went toilet. (0.4) And she was
4		up there for some ti::me. (1.0) An::d uh:: (0.3) I says,
5		'Blythe.' (0.4) 'Blythe.' (0.4) 'Blythe.' (0.6) Eventually I
6		went up. (0.7) She was sitting on the loo.=Lea::nin' (.)
7		over the s:ink. (0.3) And she was going blue.
8	Pat:	Mm-hm.
9	**Doc:**	**Oh gosh.**
10	Pat:	((Coughs))
11	Com:	Y:[ou know uh:]
12	Doc:	**[Must have been fright]ening.**
13	Com:	It was really scary. She couldn't breathe.

A traditional approach to empathy might focus on what the doctor is "feeling" at this point in the interaction. This might involve, for example, giving the doctor a questionnaire after the consultation and asking what they were thinking about when they displayed empathy. It could also involve giving such a questionnaire to the patient and asking whether they felt empathised with at this moment. (See, again, Hemmerdinger et al. (2007) and Pedersen (2009) for reviews of such measures.)

By contrast, this analysis is grounded in an understanding of what role the doctor's empathic turns at lines 9 and 12 actually play in the interaction. In other words, *why* does the doctor display empathy at this stage

[2] Extracts are presented using the conventions laid out by Jefferson (2004). These conventions were developed specifically for the transcription of talk-in-interaction, and thus render not only *what* was said, but *how* it was said (intonation, breathiness, etc.), where it was said (e.g. when two speakers talk in overlap), and any pauses or gaps within the talk. A glossary of the symbols used in this chapter can be seen in the appendix. (See, also, Hepburn and Bolden (2017) for a more detailed discussion.)

and what effect does it have on the unfolding interaction? To answer this, we can turn to the story that the patient's friend has been telling up to this point. We can see from lines 1–6 how the patient's friend has been building up to the point where she discovered the patient in the bathroom. She ultimately reaches this point at lines 6–7 where she describes how she found her friend "Lea::nin' (.) over the s:ink … going blue." This is clearly, then, a shocking emotional moment in the context of the story that she is telling. The doctor empathically recognises this emotional impact by both doing a response cry ("Oh gosh") simulating the shock that one must feel at finding a close friend in that situation (Heritage 2011), and in her use of the adjective "frightening" to suggest how the situation "Must have been."

Aside from the doctor showing an understanding of the emotional impact of the events of the story itself, there is a broader point here about the patient's friend telling this story to begin with. It is worth pointing out that, medically speaking, it is not strictly necessary for the doctor to hear this story. As noted above, the events within it happened at the beginning of the patient's illness, leading to her initial hospitalisation. Any relevant medical information from the story (about the patient's condition etc.) could thus be gleaned from the patient's medical records.

The patient's friend being given the opportunity to tell this story is thus, in itself, representative of the underlying philosophy of palliative care. In this environment, patients and their companions can tell stories that, within a stricter biomedical framework, would be deemed unnecessary. Furthermore, doctors, rather than cutting off or ignoring such "lifeworld" stories (as has traditionally been seen as the case in other medical environments—see, for example, Mishler (1984)), engage and empathise with them. This gives an initial illustration, therefore, of what is meant about empathy being a way for doctors to acknowledge patients' personhood in palliative care.

A further illustration of this point can be seen in Extract 4.2. This extract comes from a consultation in which the (male) patient (aged 57) is telling the (female) doctor about various stresses that he has been experiencing as a result of his illness. One such source of stress, and the focus of his story, has been the various problems that he has had getting the necessary adjustments made to accommodate his wheelchair. As with the previous case, the extract begins just as the patient is bringing his story to a particularly emotional point.

Extract 4.2 That's adding quite a bit of stress, is it? [P02/A: 20.48 - 21.48]

1	Pat:	S:o anyw↓ay (0.5) shih they're coming back u::m (1.7) on
2		(the eleventh I think.= Needs) to rip all that down and redo it.
3		(0.3)
4	Doc:	**Oh de:a:r.**
5	Pat:	U::m (0.5) (now) Nick has taken (the) day off.=She's gonna be
6		there.
7		(0.4)
8	Doc:	Mm.
9	Pat:	Because they're supposed to cover the bed up.=Because
10		obviously the door was shut.=So I d- I didn't see what was going
11		on.
12		(0.9)
13	Pat:	And when Nick came home in the evening (0.8) u::m (0.3) I tried
14		to sweep up the bits (). shih And Nick come back.=
15		I did see what was on the b↑ed.
16		(0.5)
17	Pat:	(There/It) was all (like) u::m metal filings all over the bed
18		and that.
19		(0.8)
20	Pat:	So: Nick went mad about that.=NOT (.) not at me,
21	Doc:	Mm.
22	Pat:	You know.=(But/At-) at them.=They (uh) they should have done it.
23		shih
24		(1.7)
25	Pat:	pt. So she's (taken/taking) the day off to u::m (0.3) make sure
26		she's there,
27		(0.7)
28	Pat:	And that.

```
29              (1.0)
30   Pat:       To make sure it's done properly this time.
31              (1.5)
32   Pat:       [(A:nd-     )]
33   Doc:       [And that's add]ing quite a bit of: s::tre:ss is it.
34   Pat:       Yea:::h.=
35   Doc:       =(Yeah./Mm.)
```

The structure of this extract is similar to Extract 4.1. From lines 1–2, the patient brings his story to a dramatic peak, revealing that the workers are going to have to come back again in the future to "rip all that down and redo it." Given that this work has been stressful for him before, the workers having to come back and do it again is clearly, in the context of his story, a highly negative development. The doctor displays understanding of this negative impact on the patient with her empathic turn at line 4: "Oh de:a:r."

In response to this first empathic display, the patient adds further details to his story, noting that his wife (Nick) will have to take the day off work the next time the workers come because, last time, the workers made a mess. After the patient has added these further details, the doctor displays her understanding again at line 33, noting that these events have "[added] quite a bit of: s::tre:ss" to the patient's situation.

As with Extract 4.1, the story that the patient is telling in this extract is not strictly necessary for the doctor to hear. It is not, after all, as though the doctor will be able to do anything to resolve or otherwise directly affect the situation that the patient is describing, this being a bureaucratic problem with the local authorities more than anything else. The patient's problem is not, in other words, "doctorable" (Heritage and Robinson 2006), and the conversation that he is having with the doctor is much closer to a "natural [conversation] between friends" (Barry et al. 2001, p. 497).

It is precisely this lack of doctorability, though, that makes this extract another strong example of a palliative care doctor engaging with a patient in a way that goes beyond what would be expected or needed within a traditional biomedical framework. By listening to and empathising with the patient's story, this doctor is, like the doctor in Extract 4.1, showing an appreciation of the patient as a person with his own unique experiences and emotions. Towards the end of the consultation, the patient does, in fact, show his appreciation at having been given the opportunity to tell his story, as can be seen in Extract 4.3.

Extract 4.3 Sometimes it's nice to just chat [P02/A: 56.05 - 56.22]

```
1  Doc:   And we'll met again next wee:k [yeah.]
2  Pat:                                   [Yea:]:h.
            (But) sometimes
3         it's nice to: (0.6)(to just) chat to an individual.
4  Doc:   Mm.
5  Pat:   You know.
```

Extracts 4.1 and 4.2 have both given a sense of how palliative care doctors can create the space for older patients to tell stories that are not strictly biomedical in nature. We have also seen how doctors can empathically engage with those stories. Despite having a wider scope than traditional biomedicine, though, palliative care is still grounded in biomedical conditions—the patients in Extracts 4.1 and 4.2 would not be in the hospice in the first place, after all, if they were not seriously physically ill. We now see how doctors go about engaging with patients' personhood while simultaneously addressing their physical conditions.

Acknowledging the Person Behind the Symptoms

Extract 4.4 comes as the (female) doctor is asking the (male) patient (aged 64) about his breathing. As the extract begins, the patient is talking about how he monitors his oxygen levels using an oximeter and how these levels can drop when he is engaged in physical activity.

Extract 4.4 That's scary, isn't it? [P23/F: 10.03 - 10.22]

```
1  Pat:  It doesn't matter what (0.4) what the oxygen's on,
2        (0.4)
3  Doc:  Yeah.=
```

4 Pat: =As soon as I start to do any wa::lkin', everything just
5 goes straight do:wn.=
6 Doc: =Okay.
7 Pat: (And) it goes down (.) right into (th-) .hhh hhh
8 Com: Low six[ties.]
9 Pat: [Low-] low s- low seventies, s:ixties,
10 Doc: Ok*h*ay.
11 (0.3)
12 Pat: I mean I've had it d- right down as low as fifties.
13 Doc: **° Gosh.° That's scary isn't [it.**
14 Pat: [And that- that's with: .hh with
15 the oxygen as w*h*- well.
16 Doc: Okay.
17 (0.7)
18 Doc: °Okay.°=
19 Pat: =But uh__
20 (1.1)
21 Doc: Okay. .hh So I'm hearing that you fee:l (0.3) probably a bit
22 frustrated really that the oxygen is- is the way that it i:s,
23 Pat: Yeah.
24 Doc: Have you- have you talked to Doctor Fairclough.=Have you talked to the oxygen tea:m about that.

In some senses, Extract 4.4 is similar to Extracts 4.1 and 4.2. It does, after all, feature the patient telling a story (between lines 1 and 12) and the doctor, at a particular point in that story (line 13), doing an empathic display ("Gosh. That's scary isn't it.") that acknowledges the emotional impact that the events of it have had on the patient. The crucial difference lies in the motivations behind those stories. In Extracts 4.1 and 4.2, as suggested earlier, patients told stories for reasons that would be superfluous in a strictly biomedical framework—doctors did not need to hear them to address their conditions.

In Extract 4.4, on the other hand, it is necessary for the patient to be telling his story because it is through that story that the doctor will be able to understand her condition and offer appropriate medical help. In this sense, the issue of the patient's personhood is even more relevant in Extract 4.4 because, as Mishler (1984) has noted, there is the possibility of a patient's symptoms being "stripped" of their context, becoming mere biomedical objects rather than events with potentially strong emotional consequences for those who are experiencing them. By showing her understanding of how the patient's oxygen levels dropping would be "scary" for him, the doctor ensures that the focus remains on him rather than on his symptoms. Again, then, we can see how empathy is a way of acknowledging the patient's personhood in palliative care.

However, this does not mean that the interaction then becomes *about* the patient's emotions. Rather, as we can see in the extract, after the doctor has displayed her empathic understanding of how "scary" the patient's symptoms are, she goes on to ask him whether he has spoken to "Doctor Fairclough" and "the oxygen tea:m" at line 24. This is because the patient's emotions here have a biomedical cause and are thus potentially amenable to biomedical remedy (rather than being psychosocial and beyond the realm of biomedicine, like in Extracts 4.1 and 4.2). The doctor here is thus acknowledging the patient's personhood while *simultaneously* addressing his biomedical symptoms.

Older Patients' Emotions and the Medical Agenda

To recap, we have now analysed three extracts. In the first two of these extracts, doctors gave older patients the space to tell stories that were not strictly "necessary" from a biomedical perspective, engaging and empathising with the details of them. In Extract 4.4, meanwhile, we saw how the doctor could show her understanding of a patient's personhood as he was describing emotionally charged physical symptoms, thus avoiding the "context-stripping approach" described by Mishler (1984, p. 164).

We now consider some examples where patients' individual emotions are not simply acknowledged in the course of palliative care, but actively clash with it in some way. To see what is meant by this, consider Extract 4.5. This extract comes from a consultation in which the (male) patient (aged 77) and the (female) doctor have been discussing what the patient would like to happen should he suddenly take a bad turn. The patient has said that he would like to be taken to the hospital only if there was a chance of returning him to his current, stable state. As the extract begins, the doctor is summarising her understanding of the patient's wish.

Extract 4.5 Not always the easiest thing to chat about [P38/G: 9.06 - 9.31]

```
1    Doc:   You feel happy (.)
2    Pat:   Y(h)e[s. ]
3    Doc:        [To-] to go in ((to the hospital)) if w- if
             we thou:ght it
4           would reverse things.= [If it was a rever]sible
             cause.
5    Pat:                          [Ye::s. Just-  ]
6           s'long as it got me back to square one °shorta
             thhing.°
7    Doc:   Okay.=
8    Pat:   =Ye[s.  ]
9    Doc:      [>Okay.<=]That's really helpful.=Thank you
10          [for   t]alking that through.
11   Pat:   [Yeah.]
12          (.)
13   Doc:   I know it's not always the easiest thi:ng [to uh ]
14   Pat:                                             [pt. No.]
15          [No.]
16   Doc:   [to ] chat about.
17   Pat:   No.
18          (0.3)
19   Doc:   (Mm./No.)=
20   Pat:   =But I can (1.4) all the f- when I- when it is
             ti:me .hh
21          all the family can be with me you see
```

When older patients enter palliative care, they are often asked to make difficult decisions about the future, and what they would like to happen as their condition continues to deteriorate. Such discussions are an important part of palliative care for older patients (see, e.g. Detering et al. (2010) on advanced care planning, and Seymour and Gott (2011) on end-of-life decision-making), ensuring that the care provided will, where possible, be in accordance with their wishes. For all their medical importance, though, there is the potential for such discussions to be difficult or upsetting for patients, touching as they do upon "dreaded issues" (Peräkylä 1995) such as declining health and, ultimately, death.

In Extract 4.5, we can see how the doctor balances discussion of these issues with recognition of this difficulty. From lines 1–11, the discussion these future planning is wrapping up, with the doctor clarifying the patient's stance (lines 1 and 3–4), and thanking him for having talked it through with her (lines 9–10). She then, at lines 13 and 16, shows her understanding that these matters are not "always the easiest thi:ng" for patients to "chat about." From a strict biomedical perspective, of course, whether a topic is "easy" to chat about or not is irrelevant. By showing her recognition of this difficulty at this point in the interaction, then, the doctor is ensuring that the patient's personhood remains in focus, even as she is carrying out a medical task.

An even stronger example of this kind of balancing act can be seen in Extract 4.6. This extract comes from a consultation in which the (female) patient (aged 64) and her (male) partner have been complaining about a doctor at the hospital who advised her to exercise. The patient found this advice offensive because she took it to imply that she must be exaggerating her illness. The problem for the (female) doctor here is that this advice, while perhaps badly delivered by the other doctor, is at its root medically sound. How then can she convey advice that is valid from a *medical* perspective, but would be difficult for the patient, as a person, to follow? Extract 4.6 exemplifies this dilemma; it begins as the doctor is explaining to the patient how exercise can help her condition.

Extract 4.6 I know that's hard [P35/F: 15.59 - 17.05]

1 Doc: What we (0.3) do kno::w (0.4) .hh (.) i::s tha**t**
2 **t**ry::i::ng to: (0.3) h:elp you <m:a:nage> the
3 breathlessne::ss: (0.4) so tha**t** (0.6) y:ou can be as active >as
4 you >>°chan.°<<=**Now I know that's ha:rd and [I know] you've got=**
5 Pat: [Mm-hm.]
6 Doc: =**lots of other stuff going on as we:ll,** .hhh (0.6) But
7 we do:: kno:w (.) that maintai:ning (.) a little bit of
8 activity:: .hh is the thing that's going to keep you
9 living (0.3)
10 Com: Lon[ger.
11 Pat: [Longer.
12 Doc: [longer.]

From lines 1–4, the doctor is explaining to the patient the importance of "[managing] the breathelessne::s" so that she can be "as active [as she can] >>°c*h*an.°<<." Immediately after saying this, however, the doctor at lines 4 and 6 acknowledges that this advice could be "ha:rd" for the patient to follow, given that she has "lots of other stuff going on as we:ll." She then returns to delivering her advice from lines 6–9 and 12, emphasising that "maintai:ning (.) a little bit of activity::" will keep the patient "living ... longer."

The doctor in this extract has walked a delicate tightrope. On the one hand, she has reiterated the same advice given to the patient by the doctor at the hospital—the advice that, on that previous occasion, the patient reported herself as having found offensive. As she has done so, however, she has also recognised how "ha:rd" it would be for the patient to follow this advice. The doctor's empathic recognition of the patient's personhood is particularly important here, therefore, because the patient has complained about the *lack* of such recognition shown by another doctor.

Conclusion

In this chapter, we have seen how palliative care doctors can display empathy at various points in their interactions with older patients. In the first section of the analysis, we saw how doctors could give older patients the space to tell stories around their conditions that were not, strictly speaking, necessary to hear in a medical environment. In the second section of the analysis, we saw how doctors could acknowledge the emotional and experiential difficulties caused by older patients' physical symptoms, whether that was finding it difficult to cough or finding it scary when their levels dropped. Finally, in the third analytic section, we saw how doctors could acknowledge older patients' emotions at points where those emotions clashed in some way with the medical agenda, whether that was a difficult discussion about planning for the future or a piece of advice that the patient had, on a previous occasion, found offensive.

Researchers have previously raised concerns about an overemphasis on the limited conceptions of "successful" ageing (Gott and Ingleton 2011), and the difficulty of taking into account older dying patients' individual preferences (Seymour and Gott 2011). This echoes both historical research on the idea of death as a "failure" (Clark 1999) and research showing that the dying process is highly individualistic, with each person bringing their own unique standards, expectations, and experiences (Goldsteen et al. 2006; Hart et al. 1998; Macnamara et al. 1994; Payne et al. 1996; Stringer 2007).

In this chapter, I have shown how empathy is a way for doctors to display an appreciation of older patients' individual qualities in end-of-life care, be that for its own sake (as in the first section of the analysis) or in the course of carrying out some medical task (as in the second and third sections). In line with palliative care's underlying philosophy (Baines 2011; Saunders et al. 1995), then, empathy is a key means via which doctors acknowledge older patients' personhood, ensuring that the care provided for them in the final stages of their lives remains focused on them as human beings rather than their illness as a detached biomedical object.

Aside from this core finding, there are several additional implications of the research outlined in this chapter. The first is that it demonstrates the benefits that close discursive analysis can have for research in the psychology of ageing. It shows, in other words, how an approach rooted in language and interaction can shed new light on matters that have not traditionally been approached in this way. While the discursive turn has had implications across all areas of psychology, there is much to suggest that it could have particular benefits for ageing research. Many issues around ageing are, after all, already rooted in language and communication. Although the specific focus of this chapter, therefore, has been on empathy and emotion, there is no reason why these other issues could not be equally amenable to, and could not equally benefit from, the same approach.

Adopting an approach rooted in actual interactions also means that the findings outlined in this chapter have practical implications for doctors working with older patients at the end of their lives. This is particularly true of the latter part of the analysis, where we saw how doctors could integrate empathy into clinical tasks to ensure that they took into account older patients' emotions and personhood. While we saw only two examples of this, there is no reason why empathy could not be integrated into a range of clinical tasks. The earlier parts of the analysis, as well, showed how doctors could either create the space for patients to disclose their emotions or empathise with patients' emotions in the course of dealing with other matters. Again, this is a finding with a potential applicability that goes far beyond the few examples of it shown here.

Acknowledgements I would like to thank the patients and doctors who participated in the study from which the data for this chapter were taken. I would also like to thank Ruth Parry for her helpful comments, Luke Feathers, Christina Faull, and LOROS hospice. Ruth Parry is funded by a National Institute for Health Research Career Development Fellowship CDF-2014-07-046. The data used in this research were collected by the "VERDIS" programme of research and training funded by The Health Foundation Insight Award RU33.

Appendix: Glossary of Transcription Symbols

:::	Colons indicate that the immediately preceding sound was elongated, with more colons meaning greater elongation.
___	Underlining indicates that the underlined sound was emphasised.
( )	Brackets indicate that the hearing is tenuous or unclear.
=	Equals sign indicates that two pieces of talk are connected.
.,	Indicate different forms of ending intonation. A full stop indicates a fall and a comma indicates a partial rise.
[ ]	Square brackets indicate overlapping talk.
(1.0)	Numbers in brackets indicate pauses, timed to the nearest tenth of a second.
(.)	Indicates a micropause, smaller than 0.2 seconds.
° °	Degree symbols indicate that talk was noticeably quiet.
.hh hh	Standalone h's indicate inbreaths (preceded by full stop) or outbreaths (no full stop).
> <	Inward-facing carets indicate that talk was noticeably fast.
< >	Outward facing carets indicate that talk was noticeably slow.

(h)	H inside brackets within a word indicates laughter.
Hhh	Italicised h within a word indicates breathiness.
Boldface	Used to draw attention to analytically relevant stretches of talk.
↑↑	Up and down arrows indicate sudden shifts in pitch.
.shih	Used to denote a sniff, e.g. when the speaker is crying (see Hepburn and Bolden 2017).

References

Baines, M. (2011). From Pioneering Days to Implementation: Lessons to Be Learnt. *European Journal of Palliative Care, 18*(5), 223–227.

Barry, C. A., Stevenson, F. A., Britten, N., Barber, N., & Bradley, C. P. (2001). Giving Voice to the Lifeworld. More Humane, More Effective Medical Care? A Qualitative Study of Doctor-Patient Communication in General Practice. *Social Science & Medicine, 53*(4), 487–505.

Bylund, C. L., & Makoul, G. (2002). Empathic Communication and Gender in the Physician–Patient Encounter. *Patient Education and Counseling, 48*(3), 207–216.

Clark, D. (1999). Cradled to the Grave? Terminal Care in the United Kingdom, 1948–67. *Mortality, 4*(3), 225–247.

Clark, D. (2007). From Margins to Centre: A Review of the History of Palliative Care in Cancer. *The Lancet Oncology, 8*(5), 430–438.

Detering, K.M., Hancock, A.D., Reade, M.C., & Silvester, W. (2010). The Impact of Advance Care Planning on End of Life in Elderly Patients: Randomised Controlled Trial. *British Medical Journal.* Advanced Online Publication. doi: https://doi.org/10.1136/bmj.c1345

Edwards, D. (1999). Emotion Discourse. *Culture & Psychology, 5*(3), 271–291.

Edwards, D., & Potter, J. (1992). *Discursive Psychology.* London: Sage Publications.

Goldsteen, M., Houtepen, R., Proot, I. M., Abu-Saad, H. H., Spreeuwenberg, C., & Widdershoven, G. (2006). What Is a Good Death? Terminally Ill Patients Dealing with Normative Expectations Around Death and Dying. *Patient Education and Counseling, 64*(1–3), 378–386.

Gott, M., & Ingleton, C. (2011). What Is Different About Dying Old? In M. Gott & C. Ingleton (Eds.), *Living with Ageing: Palliative and End of Life Care for Older People.* Oxford: Oxford University Press.

Harré, R. (2001). The Discursive Turn in Social Psychology. In D. Schiffrin, D. Tannen, & H. E. Hamilton (Eds.), *The Handbook of Discourse Analysis* (pp. 688–706). Oxford: Blackwell.
Harré, R., & Gillett, G. (1994). *The Discursive Mind*. London: Sage Publications.
Hart, B., Sainsbury, P., & Short, S. (1998). Whose Dying? A Sociological Critique of the "Good Death". *Mortality, 3*(1), 65–77.
Hemmerdinger, J. M., Stoddart, S. D. R., & Lilford, R. J. (2007). A Systematic Review of Tests of Empathy in Medicine. *BMC Medical Education, 7*(1), 24.
Hepburn, A., & Bolden, G. (2017). *Transcribing for Social Research*. London: Sage.
Hepburn, A., & Potter, J. (2007). Crying receipts: Time, Empathy, and Institutional Practice. *Research on Language and Social Interaction, 40*(1), 89–116.
Hepburn, A., & Potter, J. (2012). Crying and Crying Responses. In A. Peräkylä & M.-L. Sorjonen (Eds.), *Emotion in Interaction* (pp. 195–211). Oxford: Oxford University Press.
Heritage, J. (2011). Territories of Knowledge, Territories of Experience: Empathic Moments in Interaction. In T. Stivers, L. Mondada, & J. Steensig (Eds.), *The Morality of Knowledge in Conversation* (pp. 159–183). Cambridge: Cambridge University Press.
Heritage, J., & Lindström, A. (2012). Knowledge, Empathy, and Emotion in a Medical Encounter. In A. Peräkylä & M.-L. Sorjonen (Eds.), *Emotion in Interaction* (pp. 256–273). Oxford: Oxford University Press.
Heritage, J., & Robinson, J. (2006). Accounting for the Visit: Giving Reasons for Seeking Medical Care. In J. Heritage & D. Maynard (Eds.), *Communication in Medical Care: Interactions Between Primary Care Physicians and Patients* (pp. 48–85). Cambridge: Cambridge University Press.
Heritage, J., & Watson, D. R. (1979). Formulations as Conversational Objects. In G. Psathas (Ed.), *Everyday Language: Studies in Ethnomethodology* (pp. 123–162). London: Irvington.
Hochschild, A. R. (1979). Emotion Work, Feeling Rules, and Social Structure. *American Journal of Sociology, 85*(3), 551–575.
Hochschild, A. R. (2003). *The Managed Heart: Commercialization of Human Feeling* (20th anniversary ed.). London: University of California Press.
Hojat, M., Mangione, S., Nasca, T. J., Cohen, M. J., Gonnella, J. S., Erdmann, J. B., ... Magee, M. (2001). The Jefferson Scale of Physician Empathy: Development and Preliminary Psychometric Data. *Educational and Psychological Measurement, 61*(2), 349–365.

Jefferson, G. (1985). An Exercise in the Transcription and Analysis of Laughter. In T. Van Dijk (Ed.), *Handbook of Discourse Analysis: Discourse and Dialogue* (Vol. 3, pp. 25–34). London: Academic Press.
Jefferson, G. (1988). On the Sequential Organization of Troubles-Talk in Ordinary Conversation. *Social Problems, 35*(4), 418–441.
Jefferson, G. (2004). Glossary of Transcript Symbols with an Introduction. In G. H. Lerner (Ed.), *Conversation Analysis: Studies from the First Generation* (pp. 13–31). Amsterdam: John Benjamins Publishing Company.
Kitzinger, C. (2011). Working with Childbirth Helplines: The Contributions and Limitations of Conversation Analysis. In C. Antaki (Ed.), *Applied Conversation Analysis: Intervention and Change in Institutional Talk* (pp. 98–118). Basingstoke: Palgrave Macmillan.
Kupetz, M. (2014). Empathy Displays as Interactional Achievements—Multimodal and Sequential Aspects. *Journal of Pragmatics, 61*, 4–34.
Kuroshima, S., & Iwata, N. (2016). On displaying Empathy: Dilemma, Category, and Experience. *Research on Language and Social Interaction, 49*(2), 92–110.
McNamara, B., Waddell, C., & Colvin, M. (1994). The Institutionalization of the Good Death. *Social Science & Medicine, 39*(11), 1501–1508.
Mercer, S. W., Maxwell, M., Heaney, D., & Watt, G. C. (2004). The Consultation and Relational Empathy (CARE) Measure: Development and Preliminary Validation and Reliability of an Empathy-Based Consultation Process Measure. *Family Practice, 21*(6), 699–705.
Miller, R. J. (1992). Hospice Care as an Alternative to Euthanasia. *The Journal of Law, Medicine & Ethics, 20*(1–2), 127–132.
Mishler, E. (1984). *The Discourse of Medicine: Dialectics of Medical Interviews*. New Jersey: Ablex Publishing Corporation.
Moore, J. (2009). *Talk in Mind: The Analysis of Calls to a Mental Health Information Line* (Unpublished Doctoral Dissertation). Loughborough: Loughborough University.
Muntigl, P., & Horvath, A. O. (2014). "I Can See Some Sadness in Your Eyes": When Experiential Therapists Notice a Client's Affectual Display. *Research on Language and Social Interaction, 47*(2), 89–108.
Muntigl, P., Knight, N., & Watkins, A. (2014). Empathic Practices in Client-Centred Psychotherapies: Displaying Understanding and Affiliation with Clients. In E.-M. Graf, M. Sator, & T. Spranz-Fogasy (Eds.), *Discourses of Helping Professions* (pp. 33–58). Amsterdam: John Benjamins Publishing Company.

Payne, S. A., Langley-Evans, A., & Hillier, R. (1996). Perceptions of a 'Good' Death: A Comparative Study of the Views of Hospice Staff and Patients. *Palliative Medicine, 10*(4), 307–312.

Pedersen, R. (2009). Empirical Research on Empathy in Medicine: A Critical Review. *Patient Education and Counseling, 76*(3), 307–322.

Peräkylä, A. (1995). *AIDS Counselling: Institutional Interaction and Clinical Practice*. Cambridge: Cambridge University Press.

Peräkylä, A., & Sorjonen, M.-L. (Eds.). (2012). *Emotion in Interaction*. Oxford: Oxford University Press.

Psathas, G. (1995). *Conversation Analysis: The Study of Talk-in-Interaction*. London: Sage Publications.

Pudlinski, C. (2005). Doing Empathy and Sympathy: Caring Responses to Troubles Tellings on a Peer Support Line. *Discourse Studies, 7*(3), 267–288.

Ruusuvuori, J. (2005). "Empathy" and "Sympathy" in Action: Attending to Patients' Troubles in Finnish Homeopathic and General Practice Consultations. *Social Psychology Quarterly, 68*(3), 204–222.

Ruusuvuori, J. (2007). Managing Affect: Integration of Empathy and Problem-Solving in Health Care Encounters. *Discourse Studies, 9*(5), 597–622.

Ruusuvuori, J. (2013). Emotion, Affect and Conversation. In J. Sidnell & T. Stivers (Eds.), *The Handbook of Conversation Analysis* (pp. 330–349). Oxford: Wiley-Blackwell.

Sacks, H. (1995). *Lectures on Conversation: Volumes I & II*. Oxford: Blackwell Publishing.

Sandvik, M., Eide, H., Lind, M., Graugaard, P. K., Torper, J., & Finset, A. (2002). Analyzing Medical Dialogues: Strength and Weakness of Roter's Interaction Analysis System (RIAS). *Patient Education and Counseling, 46*(4), 235–241.

Saunders, C. (2001). The Evolution of Palliative Care. *Journal of the Royal Society of Medicine, 94*(9), 430–432.

Saunders, C., Baines, M., & Dunlop, R. (1995). *Living with Dying: A Guide to Palliative Care*. Oxford: Oxford University Press.

Selting, M. (2010). Affectivity in Conversational Storytelling: An Analysis of Displays of Anger or Indignation in Complaint Stories. *Pragmatics, 20*(2), 229–277.

Seymour, J., & Gott, M. (2011). The Challenges of Health Technology for Ageing and Dying. In M. Gott & C. Ingleton (Eds.), *Living with Ageing and Dying: Palliative and End of Life Care for Older People*. Oxford: Oxford University Press.

Silverman, D., & Peräkylä, A. (1990). AIDS Counselling: The Interactional Organisation of Talk About "Delicate" Issues. *Sociology of Health & Illness, 12*(3), 293–318.

Stone, A. L., Tai-Seale, M., Stults, C. D., Luiz, J. M., & Frankel, R. M. (2012). Three Types of Ambiguity in Coding Empathic Interactions in Primary Care Visits: Implications for Research and Practice. *Patient Education and Counseling, 89*(1), 63–68.

Stringer, S. (2007). Quality of Death: Humanisation Versus Medicalisation. *Cancer Nursing Practice, 6*(3), 23–28.

Suchman, A. L., Markakis, K., Beckman, M. B., & Frankel, R. (1997). A Model of Empathic Communication in the Medical Interview. *JAMA, 277*(8), 678–682.

Voutilainen, L. (2012). Responding to Emotion in Cognitive Psychotherapy. In A. Peräkylä & M.-L. Sorjonen (Eds.), *Emotion in Interaction* (pp. 235–255). Oxford: Oxford University Press.

Weatherall, A. (2015). "But Whose Side Are You on?" Doing Being Independent in Telephone-Mediated Dispute Resolution. In F. H. G. Chevalier & J. Moore (Eds.), *Producing and Managing Restricted Activities: Avoidance and Withholding in Institutional Interaction* (pp. 151–180). Amsterdam: John Benjamins.

Weatherall, A., & Stubbe, M. (2015). Emotions in Action: Telephone-Mediated Dispute Resolution. *British Journal of Social Psychology, 54*(2), 273–290.

Weiste, E., & Peräkylä, A. (2014). Prosody and Empathic Communication in Psychotherapy Interaction. *Psychotherapy Research, 24*(6), 687–701.

Wilkinson, S., & Kitzinger, C. (2006). Surprise as an Interactional Achievement: Reaction Tokens in Conversation. *Social Psychology Quarterly, 69*(2), 150–182.

Wispé, L. (1986). The Distinction Between Sympathy and Empathy: To Call Forth a Concept, a Word Is Needed. *Journal of Personality and Social Psychology, 50*(2), 314–321.

Wynn, R., & Wynn, M. (2006). Empathy as an Interactionally Achieved Phenomenon in Psychotherapy. *Journal of Pragmatics, 38*(9), 1385–1397.

Part II

Cognitive and Biological Psychology

5

Stress, the Immune System, and Healthy Ageing

Anna C. Whittaker

Introduction

The stress response in humans, although an adaptive mechanism initially, has the potential to be chronic and detrimental to the organism if too large and/or prolonged (Sapolsky 2007). This particularly seems to be the case later in the lifespan; in fact, although ageing is a physiological process that is part of normal development (Cutler 1991a, b), some of the changes in older age mirror the chronic effects of psychological stress on several of the body's biological systems. This chapter mainly focuses on the impact of stress on the immune system and the implications for health in older age, as stress effects on all bodily systems are beyond the scope of one chapter. Further, as the immune system undergoes several changes with ageing, this results in increased susceptibility to infectious diseases, all of which are also influenced by stress. The chapter begins

A. C. Whittaker (✉)
School of Sport, Exercise and Rehabilitation Sciences, University of Birmingham, Birmingham, UK
e-mail: a.c.phillips@bham.ac.uk

E. Peel et al. (eds.), *Psychologies of Ageing*,
https://doi.org/10.1007/978-3-319-97034-9_5

with an introduction to psychological stress and its relationship to illness susceptibility and a summary of the physiological responses to stress. It then provides an introduction to the immune system and the effects of ageing on it, including the impacts of ageing on stress hormones. Following this background, the chapter examines research evidence in this area, focusing on the potentially negative impacts of caregiver stress on immunity in ageing and the potentially positive impacts of social support.

Psychological Stress

Psychological stress can be characterised in a number of ways but is usually considered as something that exceeds our coping abilities (Lazarus and Folkman 1984). One way of characterising psychological stress in order to understand its sources is to categorise it as cataclysmic events, stressful life events, or daily hassles. Cataclysmic events include natural disasters and war which can have considerable effects on health. One example would be the impact of being a soldier during the Vietnam War, which led to the development of post-traumatic stress disorder (PTSD) in many veterans, although not all. Those veterans who developed PTSD were then at increased risk of a range of non-communicable diseases over the next 20 years compared to those who did not respond to the stress or war with PTSD (Boscarino 1997). Stressful life events are occurrences that happen to most individuals but not necessarily regularly, and their impact can be months to years such as unemployment, bereavement, marriage/divorce/separation, and caregiving for a spouse or family member. An example of the impact of this type of stressor is illustrated in a study in which individuals who had suffered more stressful life events and generally felt more stressed were more susceptible to developing an upper respiratory tract infection after exposure to a rhinovirus (such as a common cold) than those who reported fewer stressful events and lower stress levels (Cohen et al. 1991). The final category of stressful daily hassles includes a range of generally short-lived minor inconveniences such as misplacing something or being late. Such hassles are common and regular but can also relate to health outcomes, particularly in the case of a build-up of many hassles over a short time period. For example, individuals who reported higher numbers of minor daily hassles on a particular day showed increased symptoms of malaise and illness four days later (Sheffield et al. 1996). Psychological stress can also be characterised as of different durations; acute stress lasts for minutes or hours whereas chronic stress lasts for days, weeks, months, or even years, and the resultant physiological impacts on health differ considerably, as is discussed in the next section.

The Stress Response

The physiological response to psychological stress was characterised by Walter Cannon (1929) as the 'fight or flight' response. The main function of this response is to maintain bodily homoeostasis. Biologically, the key site involved in this process is the hypothalamus (Barrett 2005), a part of the brain that communicates by sending nerve impulses to other parts of the body. In this way, the hypothalamus acts within seconds and via the sympathetic nervous system stimulates the medulla of the adrenal gland to release catecholamines (adrenalin and noradrenalin) which act on receptors throughout the body to result in several effects such as increased heart rate, blood pressure, and respiration; activation of smooth muscle; and increased core temperature and pain threshold (Charmandari et al. 2005). In addition, the hypothalamus also produces chemical messengers that act more slowly and in the next minutes travel through the hypothalamic-pituitary-adrenal (HPA) axis (Sapolsky et al. 2000). Chemical messengers in this pathway include corticotrophin releasing hormone (CRH) which stimulates the anterior pituitary gland to release another hormone, adrenocorticotropic hormone (ACTH), into circulation. The target organ of ACTH is again the adrenal gland, but this time stimulates the cortical cells that synthesise and release species-specific glucocorticoids (GC) into the blood. The tight control of these GC (mainly cortisol in humans) is sustained via negative feedback that controls and terminates the release of CRH (Griffin and Ojeda 2004). Cortisol potentiates the effects of catecholamines within the body (Charmandari et al. 2005) as well as initially activating the immune system, then later working as an anti-inflammatory agent generally suppressing immune function, to prevent harmful over-activation (Munck et al. 1984). However, when stress is prolonged, chronic, or there are repeated exposures to stressful events, this puts the body out of homoeostasis for too long and dysregulation of the stress response axis can occur resulting in detrimental effects throughout the body, for example, on the cardiovascular system (McEwen 1998; Phillips 2011; Sedova et al. 2004) and also by suppression of the immune system (Sorrells and Sapolsky 2007). Acute stress can in fact be beneficial for the immune system, leading to temporary upregulation of immune cells responses, and has been associated in some cases with improved antibody responses to vaccination, for example (Edwards et al. 2006). In contrast, chronic stress, such as caregiving for a child with developmental delay, has been shown to down-regulate immunity, resulting in a decrease in the antibody response to vaccination, for example (Gallagher et al. 2009a).

The Immune System

The immune system can be considered as two distinct but interconnected elements: the innate and the adaptive immune systems. The innate response is often referred to as the 'first line of defence' against infection as it comprises mechanisms that are the first to react to an infection. The adaptive immune system is slower to respond but has the advantages that it includes memory of pathogens encountered and that its response is specific for each pathogen, thus conferring a tailored and long lasting protection against further infection by the same pathogen. The innate immune system consists of soluble components, namely the complement system, which consists of a series of proteins in the blood which, when activated, stimulate, or complement, the work of the cellular elements. The cellular elements, including neutrophils which deal with rapidly dividing bacteria and fungi, eosinophils which respond to parasitic infections, macrophages which secrete soluble factors (such as the cytokines TNFα, IL-1, IL-6) to co-ordinate and amplify the immune response and also provide immunity against intracellular bacteria, and Natural Killer (NK) cells which detect and kill virally infected cells and tumour cells.

Adaptive immunity is provided by T and B lymphocytes, which develop and mature in the thymus and bone marrow respectively. T cells can be further classified into CD4 expressing helper cells (which in turn can be split into Th1 and Th2 types), CD8 expressing cytotoxic cells, and CD25 expressing T regulatory cells, which have immune suppressive function. B cells, when presented with an antigen (by dendritic cells or with T cell help) produce antibodies to provide extended protection against infections. When a naïve T or B cell encounters a pathogen it will proliferate and differentiate into an effector cell or a memory cell, so that if the pathogen is encountered a second time a more rapid response can be achieved. Ageing is known to have deleterious effects upon both the innate and adaptive immune responses, though the latter is much better characterised (Phillips et al. 2007).

Ageing and Immunity

With ageing, the innate immune response goes through some changes. For example, complement activation appears to be unaffected, but neutrophil bactericidal and phagocytic function in vitro is dramatically reduced (Butcher et al. 2001). Macrophage function is also modified, although the literature is rather contradictory, including reports of reduced capacity to engulf pathogens, release chemicals such as superoxides to kill pathogens, as seen in neutrophils, but enhanced secretion of the cytokines IL-6 and IL-8 in response to pathogens. NK cells are also affected by ageing; while their numbers do not change with age, their cytotoxic capacity is reduced (Hazeldine et al. 2012) which has also been shown to relate to reduced survival in people aged over 75 years (Ogata et al. 2001). In the adaptive immune system, the thymus gland atrophies and thus fewer naïve T cells are produced. As the size of the T cell pool is maintained at a constant level, the proportion of T cells that are memory cells increases. Consequently, as we age, we are less able to deal with new pathogens. In addition to changes in the ratio of naïve to memory T cells, there is a shift from T-helper 1 to T-helper 2 cells, and the end result is reduced cell-mediated, Th1-type, immunity. Finally, with ageing, antibody production in response to antigen declines; for example, older people produce a lower antibody titre (a measure of how much antibody a person has produced) in response to vaccination than younger individuals and the antibodies produced are of lower affinity (binds less well to the invading antigens). This is thought to be largely the result of a decline in T cell help for B cells in older adults.

Another concept that frequently appears in the literature when discussing the ageing of the immune system is inflammaging (Franceschi et al. 2007). Inflammatory factors are damaging in excessive amounts and inflammaging indicates an imbalance between inflammatory factors necessary to fight the infection and anti-inflammatory components that act as a counter weight. It has been suggested that ageing and longevity could, therefore, potentially be dependent on this balance (Franceschi et al. 2007). This would mean that immunosenescence, together with inflammatory markers such as different cytokines (IL-6, IL-8, and IL-15), could contribute to the prediction predictors of the longevity of organisms.

Stress Hormones and Ageing

As outlined earlier, stress, whether physical or psychological, is broadly sensed by two systems within the hypothalamus, the HPA axis and the sympathetic-adrenal-medullary system. Stress induces the release of catecholamines from the adrenal medulla and both cortisol and dehydroepiandrosterone (DHEA) from the adrenal cortex. Catecholamines and cortisol can both be immunosuppressive if chronically elevated. In contrast, DHEA is a precursor to sex hormones and is considered to be immune enhancing (Butcher et al. 2005). Due to the impact of these hormones on immunity, any change in their production could therefore have significant health implications. In humans, the production of DHEA and its sulphated form, DHEAS, declines with age, a process termed the adrenopause (Orentreich et al. 1984). The synthesis of DHEA is maximal in humans at age 20–30 and declines gradually thereafter, so that by the seventh decade levels of DHEA can be as low as 10% of that seen in young adulthood (Orentreich et al. 1992); this adrenopause occurs at similar rates in both males and females. However, although DHEA/S levels fall with age, the production of glucocorticoids such as cortisol is remarkably unaltered (Orentreich et al. 1992), resulting in a relative excess of cortisol over DHEA/S and an imbalance of immune suppression over immune enhancement. The age-related immunological and endocrinological changes outlined earlier may have implications for the ability to cope (physiologically) with stress in older adults. It is likely that the combination of adrenopause, leading to a relative preponderance of cortisol, and an already reduced immune defence against infection through immune senescence, may leave this population particularly vulnerable to the negative effects of stress on immunity (Graham et al. 2006; Phillips et al. 2007). However, younger adults are also susceptible to the impact of stress on immunity. For example, various studies have now shown reduced antibody responses to vaccination in those with greater self-reported stress levels (Burns et al. 2003; Phillips et al. 2005). Nonetheless, when ageing is combined with stress, the effects on immunity are often more detrimental. For example, in one large study, caregivers, but only those aged over 60 years, showed lower levels of a particular antibody, salivary immunoglobulin A, which targets pathogens in biological fluids, particularly saliva at mucosal surfaces (Gallagher et al. 2008a).

Caregiver Stress and Immunity in Ageing

As indicated earlier, one commonly studied model of the impact of stress on immunity is the role of caregiving for another person, be it a spouse or child with a physical or mental illness or disability. Older caregivers have most commonly been studied in this context, using the model of family dementia caregiving (Gouin et al. 2008). Caregiving is now well established as having a serious effect on psychological wellbeing and physical health among caregivers when compared to matched non-caregiving individuals (Pinquart and Sorensen 2003). Both innate and adaptive immunity are affected by chronic stress experienced by older adults. For example, wound healing was slower in older dementia caregivers when compared to age, sex, and income-matched controls (Kiecolt-Glaser et al. 1995). Further, lower production of proinflammatory cytokines involved in the wound healing process such as IL-1α, IL-8 (Glaser et al. 1999), as well as IL-1β (Kiecolt-Glaser et al. 1995) has been observed in caregivers compared to controls. NK cells are essential in targeting tumour or virally infected cells. (Esterling et al. (1994) showed that NK cells from caregivers respond more weakly compared to NK cells from controls.)

A further association with the chronic stress of caregiving was found for adaptive cell-mediated immunity; elevated cortisol levels as well as poorer proliferation to antigen and lower IL-2 production was shown in a caregiving group (Bauer et al. 2000). Caregiving stress in older adults has also been shown to be associated with the T-helper 1 to T-helper 2 shift in the type of cytokine responses, with the difference that in older stressed individuals this was driven purely by an increase in IL-10 production, with no difference in IFN-γ production by Th1 cells (Glaser et al. 2001). It is likely that stress-induced changes in catecholamine levels (Elenkov and Chrousos 2002) during the psychological stress response drive this cytokine-related behaviour.

Inflammaging, as observed in older adults, might also be more severe among chronically stressed older adults, such as dementia caregivers. Indeed, when compared to non-caregiving older adults who also had immunosenescence, not only did older caregivers show higher levels of IL-6 (von Kanel et al. 2006) but its rate of increase was four times higher than in non-caregiving older controls, leaving them particularly vulnerable to IL-6 related diseases such as frailty, cardiovascular diseases, osteoporosis, and others (Ershler and Keller 2000).

A novel approach for assessing the severity by which caregiving stress affects the immune system of older caregivers is that of studies of latent-virus antibody titres. It is known, for example, that reactivation of latent viral infections, such as those initiated by the Herpes group (HSV-1, EBV, and CMV) is typical for immunosuppressed patients such as HIV and transplant patients (Rasmussen 1991). Interestingly, older spousal caregivers had higher IgG antibody titres against EBV VCA (virus capsid antigen) compared to the matched controls, indicating poorer control of the latent infection in this group (Kiecolt-Glaser et al. 1991). Together with the higher antibody titre to total viral antigen of HSV-1, caregivers also had a decreased virus-specific T cell response; another component of immune system necessary for controlling the infection (Glaser and Kiecolt-Glaser 1997). Older caregivers have also been characterised by higher antibody titres against CMV when compared to the controls (Pariante et al. 1997). In this instance, showing higher antibody levels against CMV is an indicator that the virus is not under control and therefore indicates poorer immunity.

Vaccination responses are affected by increasing age which makes older adults particularly vulnerable to frequent infections such as pneumonia and influenza, among the top five causes of high morbidity and mortality in this age group (Thompson et al. 2003). It would be expected that this aspect of immune incompetence would be further exacerbated in older adults affected by the chronic stress of caregiving. This is indeed the case; a significantly lower percentage of older caregivers of dementia patients showed a four-fold increase in antibody titre in response to vaccination against the influenza virus, a response that is clinically considered to be protective against infection (Vedhara et al. 1999a). This was accompanied by higher salivary cortisol concentration in the caregiver group when compared to the controls, pointing again to the role of HPA axis in immune regulation among chronically stressed individuals. Most antigens, however, trigger both humoral, that is, the antibody response which is generated by B lymphocytes, as well as cellular responses, mainly mediated by cytotoxic CD8+ T cells (Glaser et al. 2000; Kiecolt-Glaser et al. 1996; Siergist 2008). In addition, CD4+ helper T cells are necessary as mediators between those two. It has been shown that both the antibody response to medical vaccination against the influenza virus, as well as IL-2 production in response to antigen stimulation, was lower in caregivers comparing to the controls (Kiecolt-Glaser et al. 1996). In the case of the pneumococcal pneumonia vaccine, even though caregivers managed to exert an adequate immune response initially, shown as a rise in IgG antibody titre, it declined over time more rapidly in this group than in the group of matched controls, likely either as a consequence of decrease in number of antibody-specific B cells, or their ability to produce antibody (Glaser et al. 2000; Vedhara et al. 1999a). However, this impact of caregiving stress on vaccination responses is not unique to older caregivers as was previously thought (Vedhara et al. 2002). One elegant study showed that if the caregiving stress is severe enough, particularly if the behaviours exhibited by the care recipient are challenging, then even younger caregivers can display decrements in the antibody response to vaccination. This was shown in young parental caregivers of children with a learning delay such as autism, Down's Syndrome, and other less well-known syndromes, in comparison to parents of typically developing children. Parental caregivers showed a reduction in the production of antibodies to the influenza and pneumococcal vaccinations, and this was particularly marked among those whose children displayed severe challenging behaviours (Gallagher et al. 2009a, b). However, it should be noted that there was a trend for an effect for ageing such that the lowest antibody titres were among parental caregivers at the older end of the age group (Gallagher et al. 2009a). These studies have been followed up with similar case-control design studies but including both younger and older adults to examine the interaction between ageing and caregiving stress. These studies showed that for some aspects of immunity, ageing or the interaction between stress and ageing were not the key predictors of poorer immune function, but the stress itself was. In these studies caregivers reporting higher psychological distress (depression, anxiety, and caregiving burden) showed the poorest neutrophil function independent of age (Vitlic et al. 2015), and individuals with higher levels of psychological distress generally also displayed a somewhat higher cortisol:DHEAS ratio within each age group (Vitlic et al. unpublished data).

Chronic Stress and Immunity in Ageing

Given the importance of the caregiving model of stress, little attention has been given to other sources of stress among older adults and immunity. Although one study has reported that perceived stress, measured using the perceived stress questionnaire (Cohen et al. 1983), was associated with a poorer antibody response to the influenza vaccine in older adults (Kohut et al. 2002), another small-scale study found no association between perceived stress and antibody status following this vaccination in very old (mean age 84 years) nursing home residents (Moynihan et al. 2004). Very few studies have focused on stressful life events, despite these being a common means of assessing the impact of stress on immunity in younger samples (e.g., Burns et al. 2003; Phillips et al. 2005). However, a study published in 2006 examined overall stressful life events using a life events rating scale and showed that middle-aged and older adults with higher ratings of stress and disruptiveness for the stressful events they had experienced in the past two years showed lower levels of IgA in saliva (Phillips et al. 2006b). Further, another study took advantage of the UK National Health Service vaccination programme whereby adults aged 65+ years are invited for an annual influenza vaccination. Stressful life events and social support were measured using standardised psychometric questionnaires and antibody titres were assessed at baseline prior to vaccination and one month later. Although there was no overall effect of stress and social support, one particular stressor of import to older adults related to their immune responses; those who had suffered bereavement in the past year showed lower antibody titres to two of the three 'flu vaccine components' (Phillips et al. 2006a). The absence of an association between overall life events and antibody response to influenza vaccination in the older adults and vaccination study (Phillips et al. 2006a) contrasts with the results of previous research on young participants (Burns et al. 2003; Phillips et al. 2005). However, in these student studies, the modal number of life events experienced in the past year was six, with no participants reporting one or less events (Phillips et al. 2005), whereas in the older sample, the modal number of major life events in the year prior to vaccination was zero, with 31% of the sample reporting no events and a further 17% reporting only one. This might be due to differences between student and older adult life event stress scales. In the student studies, less serious events were included in the stress scale, for example, getting an unjustified low mark on a test or minor financial problems, along with more major events, whereas the older adults' life events scale tended to focus on exposure to major life events. Accordingly, the absence of an association between antibody response and overall life events in older adults may reflect the use of a scale including only serious, rarer life events. However, the results for bereavement would argue against this explanation. In addition, it is also possible that older people simply experience fewer general life events than younger samples. There is certainly evidence to this effect: older individuals encountered fewer major life events than middle-aged participants in a large cohort study in the west of Scotland using the same life events measure as the Phillips et al. (2006a, b) study, but retrospectively over two years. Middle-aged (mean age 44 years) participants identified a mean of 2.0 events whereas the mean number of events for the older group (mean age 63 years) was 1.7 (Carroll et al. 2005). These data also suggest that the Phillips et al. (2006a, b) participants were not unusual in experiencing few life events, given that the mean number of events reported over one year was 2.9. Accordingly, it may be that individual differences in general life events exposure are less important for immunity as people age, whereas bereavement, a specific life event that older people are more likely to encounter than the young, assumes greater prominence.

The negative association between bereavement and antibody status following vaccination is in line with previous studies of bereavement and immune function. Bereavement has been associated with in vitro functional immune measures such as decreased natural killer cell cytotoxicity and poorer lymphocyte proliferation to antigen (Bartrop et al. 1977; Goodkin et al. 1996; Irwin et al. 1987; Kemeny et al. 1995; Schleifer et al. 1983; Zisook et al. 1994). In follow-up work, focussing on the two-month period post-bereavement, it was shown that neutrophils' killing ability was suppressed in bereaved older adults, an effect that was accompanied by the increase in cortisol:DHEAS ratio (Khanfer et al. 2011). However, when this study was replicated and also included younger adults, the impact of bereavement stress on neutrophil function was only evident for older caregivers (Vitlic et al. 2014). Previous work has suggested that psychological morbidity may be a mediator between the bereavement and both immune system effects and their impact on risk of morbidities such as influenza and pneumonia. For example, Zisook et al. (1994) reported an effect on immune indices where the bereaved older adults were reporting more depressive symptoms Other work has indicated that the influence of conjugal bereavement on mortality is modified by gender, existing cardiovascular disease status, and depression (Stahl et al. 2016).

Stress need not be only psychological stress but could also be physiological stress such as physical disease or physical trauma such as a severe fracture or burn. Older adults who experienced the physical trauma or stress of hip fracture had higher cortisol:DHEA ratios than healthy controls and lower neutrophil function (Butcher et al. 2005), and individuals with lower neutrophil function were more likely to succumb to infection post-fracture (Butcher et al. 2003). Further, older adults who developed depression post-hip fracture showed the highest cortisol:DHEA ratio and poorest neutrophil function (Duggal et al. 2013), as well as worse frailty and slower physical recovery (Phillips et al. 2013).

Social Support and Immunity in Ageing

Given the substantial impact of various types of stress on immunity in older adults, as outlined earlier, understanding psychological factors that can help to enhance or improve immunity in this group is particularly important. Social support, or comfort, caring, esteem, or help provided by other people or social groups can be a key resource that helps individuals cope with life. It has also been shown to have a substantial impact on health; for example, individuals with low numbers of supportive relationships had two to three times the mortality risk compared to those with large social networks (Berkman and Syme 1979). Indeed, social network size and quality and frequency of social support have been shown to impact on morbidity and mortality from serious diseases in many epidemiological studies, for example (Barger 2013; House et al. 1982; Kaplan et al. 1988). Social support has also been shown to relate to immune function. For example, whereas students who had seroconverted (developed antibodies) after the first injection of the standard three-dose hepatitis B vaccination were less anxious and reported lower stress levels than those who had not, those who reported greater social support demonstrated a stronger combined immune response to the booster third inoculation (Glaser et al. 1992). In another study of college first-year students, loneliness and smaller social network size were associated with a poorer antibody response to the A/New Caledonian strain of the influenza vaccination (Pressman et al. 2005). Finally, higher social support scores, particularly higher frequency of tangible support, were related to an increased antibody response to the A/Panama component of the influenza vaccination, again in university students (Phillips et al. 2005).

In a study of social support in older adults, social support was negatively correlated with A/Panama influenza strain antibody status following vaccination, a finding which contradicts the studies with students above that even the authors found difficult to explain (Moynihan et al. 2004). In contrast, a larger study of older adults considered the actual vaccination response, that is, the change in antibody levels from pre- to post-vaccination (Phillips et al. 2006a). In this study, although social network size and functional social support were not related to antibody response, married/cohabiting participants showed a better antibody response to the A/Panama strain at one month than those who were not married, particularly widowed, participants. Also, for those who were married or cohabiting, higher marital satisfaction was related to higher titres to A/Panama at one month. This is not entirely surprising given that poorer marital quality, in terms of adjustment and negative marital interactions, is associated with inferior functional immunity evidenced through reduced proliferation to some antigens, poorer latent virus control (Kiecolt-Glaser et al. 1987, 1997, 1988, 1993), and weaker NK cell cytotoxicity (Miller et al. 1999) in the general population. Further, it is possible that the variations between older caregivers and controls in terms of vaccination response (Glaser et al. 1998, 2000; Kiecolt-Glaser et al. 1996; Vedhara et al. 1999b) may be driven, at least in part, by the effects of caregiving on marital quality and satisfaction, although more specific measurement of stressful life events and marital parameters would be necessary to support this speculation. Whatever the case, these findings resonate with the broad consensus that both marriage (Gordon and Rosenthal 1995; House et al. 1988; Johnson et al. 2000; Verbrugge 1979) and marital satisfaction (Coyne and DeLongis 1986; Kiecolt-Glaser and Newton 2001; Robles and Kiecolt-Glaser 2003) are beneficial for health. Further, it is possible that in an older aged population, general social support is less critical, whereas the specific social support resource of a happy marriage becomes more important for health, including susceptibility to infection.

Interestingly, the studies of social support and immunity show direct associations rather than an effect of social support via buffering the negative impact of stress. It is possible that for psychological and other health outcomes, social support can buffer stress effects (Lazarus and Folkman

1984; Rosengren et al. 1993), whereas for immune function, social support might impact immunity independently. Certainly, social support has been shown to affect different types of vaccination responses to those impacted by stressful life events in various studies among younger adults (Gallagher et al. 2007, 2008b; Phillips et al. 2005).

Conclusion

This chapter has focused on the impact of ageing on stress hormones and immune function as well as the effects of stress on immunity and the role of stress hormones. It also briefly touched on a variable that is known to relate to stress often, social support, and its links with immune function in ageing. Ageing is considered as a normal process, but in some cases the normal processes of ageing (adrenopause, inflammaging, immune senescence) can contribute to a reduced ability to deal physiologically with stress in later life. However, the research has also shown that stress can impact on stress hormone levels and certain aspects of immunity even in young healthy adults; thus, chronic stress throughout the life span will likely further impact on health and wellbeing as these individuals age. Consequently, individuals with a life history of fewer severe stress exposures may be at lower risk of a heightened cortisol:DHEA ratio, immune decrements, and greater inflammation, even in the presence of the normal hormonal and immune changes associated with ageing. These individuals are thus likely to be more resilient to stress or trauma if and when it does occur later in the lifespan, although longitudinal research would be needed to confirm this possibility.

Although avoiding stressful events themselves may not be a realistic undertaking for most individuals, certainly where stress levels can be reduced by healthy behaviours or seeking social support, these methods are likely to have positive psychological, immune, and thus health impact throughout life, not just in older age. Healthy behaviours with direct effects on both perceived stress levels and immune function can also be pursued in order to increase resilience in later life. These would include exercise or physical activity, adequate sleep, a balanced diet, not smoking or taking drugs, and moderation of alcohol intake, some of which is outlined in Chapter 6, but their effects on immunity and within healthy ageing warrant a separate chapter each.

A certain level of stress can be beneficial for health and indeed the immune system. As described in this chapter, this occurs in the case of acute stress, where the immune system can demonstrate enhanced function in response in much the same way a vaccine challenges the immune system (Lewitus and Schwartz 2009). The challenge for healthy ageing is for the occurrence of acute stressors not to escalate to the extent where they lead to overload and individuals beginning to show the detrimental physiological effects of chronic stress.

References

Barger, S. D. (2013). Social Integration, Social Support and Mortality in the US National Health Interview Survey. *Psychosomatic Medicine, 75*(5), 510–517. doi:PSY.0b013e318292ad99 [pii]https://doi.org/10.1097/PSY.0b013e318292ad99

Barrett, E. J. (2005). The Adrenal Gland. In W. F. Boron & E. L. Boulpaep (Eds.), *Medical Physiology: A Cellular and Molecular Approach* (pp. 1049–1065). New York/Amsterdam: Elsevier Inc.

Bartrop, R. W., Luckhurst, E., Lazarus, L., Kiloh, L. G., & Penny, R. (1977). Depressed Lymphocyte Function After Bereavement. *Lancet, 1*(8016), 834–836.

Bauer, M. E., Vedhara, K., Perks, P., Wilcock, G. K., Lightman, S. L., & Shanks, N. (2000). Chronic Stress in Caregivers of Dementia Patients Is Associated with Reduced Lymphocyte Sensitivity to Glucocorticoids. *Journal of Neuroimmunology, 103*(1), 84–92. https://doi.org/10.1016/s0165-5728(99)00228-3

Berkman, L. F., & Syme, S. L. (1979). Social Networks, Host Resistance, and Mortality: A Nine-Year Follow-Up Study of Alameda County Residents. *American Journal of Epidemiology, 109*(2), 186–204.

Boscarino, J. A. (1997). Diseases Among Men 20 Years After Exposure to Severe Stress: Implications for Clinical Research and Medical Care. *Psychosomatic Medicine, 59*(6), 605–614.

Burns, V. E., Carroll, D., Drayson, M., Whitham, M., & Ring, C. (2003). Life Events, Perceived Stress and Antibody Response to Influenza Vaccination in Young Healthy Adults. *Journal of Psychosomatic Research, 55*, 569–572.

Butcher, S. K., Chahal, H., Nayak, L., Sinclair, A., Henriquez, N. V., Sapey, E., et al. (2001). Senescence in Innate Immune Responses: Reduced Neutrophil Phagocytic Capacity and CD16 Expression in Elderly Humans. *Journal of Leukocyte Biology, 70*(6), 881–886.

Butcher, S. K., Killampalli, V., Chahal, H., Kaya Alpar, E., & Lord, J. M. (2003). Effect of Age on Susceptibility to Post-Traumatic Infection in the Elderly. *Biochemical Society Transactions, 31*(2), 449–451.

Butcher, S. K., Killampalli, V., Lascelles, D., Wang, K., Alpar, E. K., & Lord, J. M. (2005). Raised Cortisol:DHEAS Ratios in the Elderly After Injury: Potential Impact Upon Neutrophil Function and Immunity. *Aging Cell, 4*(6), 319–324.

Cannon, W. B. (1929). *Bodily Changes in Pain, Hunger, Fear, and Rage*. New York: D. Appleton & Co.

Carroll, D., Phillips, A. C., Ring, C., Der, G., & Hunt, K. (2005). Life Events and Hemodynamic Stress Reactivity in the Middle-Aged and Elderly. *Psychophysiology, 42*, 269–276.

Charmandari, E., Tsigos, C., & Chrousos, G. (2005). Endocrinology of the Stress Response. *Annual Review of Physiology, 67*, 259–284.

Cohen, S., Kamarck, T., & Mermelstein, R. (1983). A global Measure of Perceived Stress. *Journal of Health and Social Behavior, 24*, 385–396. https://doi.org/10.2307/2136404

Cohen, S., Tyrell, D. A. J., & Smith, A. P. (1991). Psychological Stress and Susceptibility to the Common Cold. *The New England Journal of Medicine, 325*, 606–612.

Coyne, J. C., & DeLongis, A. (1986). Going Beyond Social Support: The Role of Social Relationships in Adaptation. *Journal of Consulting and Clinical Psychology, 54*(4), 454–460.

Cutler, R. G. (1991). Antioxidants and Aging. *The American Journal of Clinical Nutrition, 53*(1 Suppl), 373S–379S.

Duggal, N. A., Upton, J., Phillips, A. C., & Lord, J. M. (2013). Depression Is Associated with Reduced Neutrophil Function in Hip Fracture Patients. *Brain, Behavior, and Immunity, 33*, 173–182. https://doi.org/10.1016/j.bbi.2013.07.004

Edwards, K. M., Burns, V. E., Reynolds, T., Carroll, D., Drayson, M., & Ring, C. (2006). Acute Stress Exposure Prior to Influenza Vaccination Enhances Antibody Response in Women. *Brain, Behavior, and Immunity, 20*, 159–168.

Elenkov, I. J., & Chrousos, G. P. (2002). Stress Hormones, Proinflammatory and Antiinflammatory Cytokines, and Autoimmunity. In M. Cutolo, J. W.

J. Bijlsma, R. G. Lahita, A. T. Masi, R. H. Straub, & H. L. Bradlow (Eds.), *Neuroendocrine Immune Basis of the Rheumatic Diseases Ii, Proceedings* (Vol. 966, pp. 290–303). New York: New York Academy of Science.

Ershler, W. B., & Keller, E. T. (2000). Age-Associated Increased Interleukin-6 Gene Expression, Late-life Diseases, and Frailty. *Annual Review of Medicine, 51*, 245–270. https://doi.org/10.1146/annurev.med.51.1.245

Esterling, B. A., Kiecolt-Glaser, J. K., Bodnar, J. C., & Glaser, R. (1994). CHRONIC Stress, Social Support and Persistent Alterations in the Natural-Killer Cell Response to Cytokines in Older Adults. *Health Psychology, 13*(4), 291–298. https://doi.org/10.1037//0278-6133.13.4.291

Franceschi, C., Capri, M., Monti, D., Giunta, S., Olivieri, F., Sevini, F., et al. (2007). Inflammaging and Anti-Inflammaging: A Systemic Perspective on Aging and Longevity Emerged from Studies in humans. *Mechanisms of Ageing and Development, 128*(1), 92–105. S0047-6374(06)00249-1.[pii]10.1016/j.mad.2006.11.016

Gallagher, S., Carroll, D., & Phillips, A. C. (2007). Life Events Stress and Antibody Response to Hepatitis A Vaccination. *Brain, Behavior, and Immunity, 21* e13. Abstract.

Gallagher, S., Phillips, A. C., Evans, P., Der, G., Hunt, K., & Carroll, D. (2008a). Caregiving Is Associated with Low Secretion Rates of Immunoglobulin A in Saliva. *Brain, Behavior, and Immunity, 22*(4), 565–572. S0889-1591(07)00309-1 [pii]10.1016/j.bbi.2007.11.007

Gallagher, S., Phillips, A. C., Ferraro, A. J., Drayson, M. T., & Carroll, D. (2008b). Psychosocial Factors Are Associated with the Antibody Response to Both Thymus-Dependent and Thymus-Independent Vaccines. *Brain Behaviour & Immunity, 22*, 456–460.

Gallagher, S., Phillips, A. C., Drayson, M. T., & Carroll, D. (2009a). Caregiving for Children with Developmental Disabilities Is Associated with a Poor Antibody Response to Influenza Vaccination. *Psychosomatic Medicine, 71*(3), 341–344. https://doi.org/10.1097/PSY.0b013e31819d1910

Gallagher, S., Phillips, A. C., Drayson, M. T., & Carroll, D. (2009b). Parental Caregivers of Children with Developmental Disabilities Mount a Poor Antibody Response to Pneumococcal Vaccination. *Brain, Behavior, and Immunity, 23*, 338–346. https://doi.org/10.1016/j.bbi.2008.05.006

Glaser, R., & Kiecolt-Glaser, J. K. (1997). Chronic Stress Modulates the Virus-Specific Immune Response to Latent Herpes Simplex Virus Type 1. *Annals of Behavioral Medicine, 19*(2), 78–82.

Glaser, R., Kiecolt-Glaser, J. K., Bonneau, R. H., Malarkey, W., Kennedy, S., & Hughes, J. (1992). Stress-Induced Modulation of the Immune Response to Recombinant Hepatitis B Vaccine. *Psychosomatic Medicine, 54*, 22–29.

Glaser, R., Kiecolt-Glaser, J. K., Malarkey, W., & Sheridan, J. F. (1998). The Influence of Psychological Stress on the Immune Response to Vaccines. *Annals of the New York Academy of Sciences, 840*, 649–655.

Glaser, R., Kiecolt-Glaser, P. K., Marucha, P. T., MacCallum, R. C., Laskowski, B. F., & Malarkey, W. B. (1999). Stress-Related Changes in Proinflammatory Cytokine Production in Wounds. *Archives of General Psychiatry, 56*(5), 450–456. https://doi.org/10.1001/archpsyc.56.5.450

Glaser, R., Sheridan, J., Malarkey, W. B., MacCallum, R. C., & Kiecolt-Glaser, J. K. (2000). Chronic Stress Modulates the Immune Response to a Pneumococcal Pneumonia Vaccine. *Psychosomatic Medicine, 62*(6), 804–807.

Glaser, R., MacCallum, R. C., Laskowski, B. F., Malarkey, W. B., Sheridan, J. F., & Kiecolt-Glaser, J. K. (2001). Evidence for a Shift in the Th-1 to Th-2 Cytokine Response Associated with Chronic Stress and Aging (vol 56, pg 477, 2001). *Journals of Gerontology Series a-Biological Sciences and Medical Sciences, 56*(11), M673–M673.

Goodkin, K., Feaster, D. J., Tuttle, R., Blaney, N. T., Kumar, M., Baum, M. K., et al. (1996). Bereavement Is Associated with Time-Dependent Decrements in Cellular Immune Function in Asymptomatic Human Immunodeficiency Virus Type-1 Seropositive Homosexual Men. *Clinical and Diagnostic Laboratory Immunology, 3*, 109–118.

Gordon, H. S., & Rosenthal, G. E. (1995). Impact of Marital Status on Outcomes in Hospitalized Patients. Evidence from an Academic Medical Center. *Archives of Internal Medicine, 155*(22), 2465–2471.

Gouin, J. P., Hantsoo, L., & Kiecolt-Glaser, J. K. (2008). Immune Dysregulation and Chronic Stress Among Older Adults: A Review. *Neuroimmunomodulation, 15*(4–6), 251–259. https://doi.org/10.1159/000156468

Graham, J. E., Christian, L. M., & Kiecolt-Glaser, J. K. (2006). Stress, Age, and Immune Function: Toward a Lifespan Approach. *Journal of Behavioral Medicine, 29*(4), 389–400.

Griffin, J. E., & Ojeda, S. R. (2004). *Textbook of Endocrine Physiology*. New York: Oxford University Press.

Hazeldine, J., Hampson, P., & Lord, J. M. (2012). Reduced Release and Binding of Perforin at the Immunological Synapse Underlies the Age-Related Decline in Natural Killer Cell Cytotoxicity. *Aging Cell*, 1–9.

House, J. S., Robbins, C., & Metzner, H. L. (1982). The Association of Social Relationships and Activities with Mortality: Prospective Evidence from the Tecumseh Community Health Study. *American Journal of Epidemiology, 116*(1), 123–140.

House, J. S., Landis, K. R., & Umberson, D. (1988). Social Relationships and Health. *Science, 241*(4865), 540–545.

Irwin, M., Daniels, M., Smith, T. L., Bloom, E., & Weiner, H. (1987). Impaired Natural Killer Cell Activity During Bereavement. *Brain, Behavior, and Immunity, 1*(1), 98–104. https://doi.org/10.1016/0889-1591(87)90011-0

Johnson, N. J., Backlund, E., Sorlie, P. D., & Loveless, C. A. (2000). Marital Status and Mortality: The National Longitudinal Mortality Study. *Annals of Epidemiology, 10*(4), 224–238.

Kaplan, G. A., Salonen, J. T., Cohen, R. D., Brand, R. J., Syme, S. L., & Puska, P. (1988). Social Connections and Mortality from All Causes and from Cardiovascular Disease: Prospective Evidence from Eastern Finland. *American Journal of Epidemiology, 128*(2), 370–380.

Kemeny, M. E., Weiner, H., Duran, R., Taylor, S. E., Visscher, B., & Fahey, J. L. (1995). Immune System Changes After the Death of a Partner in HIV-Positive Gay Men. *Psychosomatic Medicine, 57*(6), 547–554.

Khanfer, R., Lord, J. M., & Phillips, A. C. (2011). Neutrophil Function and Cortisol:DHEAS Ratio in Bereaved Older Adults. *Brain, Behavior, and Immunity, 25*(6), 1182–1186. https://doi.org/10.1016/j.bbi.2011.03.008

Kiecolt-Glaser, J. K., & Newton, T. L. (2001). Marriage and Health: His and Hers. *Psychological Bulletin, 127*(4), 472–503.

Kiecolt-Glaser, J. K., Fisher, L. D., Ogrocki, P., Stout, J. C., Speicher, C. E., & Glaser, R. (1987). Marital Quality, Marital Disruption, and Immune Function. *Psychosomatic Medicine, 49*(1), 13–34.

Kiecolt-Glaser, J. K., Kennedy, S., Malkoff, S., Fisher, L., Speicher, C. E., & Glaser, R. (1988). Marital Discord and Immunity in Males. *Psychosomatic Medicine, 50*(3), 213–229.

Kiecolt-Glaser, J. K., Speicher, C. E., Glaser, R., Dura, J. R., & Trask, O. J. (1991). Spousal Caregivers of Dementia Victims—Longitudinal Changes in Immunity and Health. *Psychosomatic Medicine, 53*(4), 345–362.

Kiecolt-Glaser, J. K., Malarkey, W., Chee, M., Newton, T., Cacioppo, J. T., Mao, H. Y., & Glaser, R. (1993). Negative Behavior During Marital Conflict Is Associated with Immunological Down-Regulation. *Psychosomatic Medicine, 55*, 395–409.

Kiecolt-Glaser, J. K., Marucha, P. T., Malarkey, W. B., Mercado, A. M., & Glaser, R. (1995). Slowing of Wound-Healing by Psychological Stress. *Lancet, 346*(8984), 1194–1196.

Kiecolt-Glaser, J. K., Glaser, R., Gravenstein, S., Malarkey, W. B., & Sheridan, J. (1996). Chronic Stress Alters the Immune Response to Influenza Virus Vaccine in Older Adults. *Proceedings of the National Academy of Sciences of the United States of America, 93*(7), 3043–3047. https://doi.org/10.1073/pnas.93.7.3043

Kiecolt-Glaser, J. K., Glaser, R., Cacioppo, J. T., MacCallum, R. C., Snydersmith, M., Kim, C., & Malarkey, W. B. (1997). Marital Conflict in Older Adults: Endocrinological and Immunological Correlates. *Psychosomatic Medicine, 59*(4), 339–349.

Kohut, M. L., Cooper, M. M., Nickolaus, M. S., Russell, D. R., & Cunnick, J. E. (2002). Exercise and Psychosocial Factors Modulate Immunity to Influenza Vaccine in Elderly Individuals. *Journal of Gerontology, 57*, 557–562.

Lazarus, R. S., & Folkman, S. (1984). *Stress, Coping and Adaptation*. New York: Springer.

Lewitus, G. M., & Schwartz, M. (2009). Behavioral Immunization: Immunity to Self-Antigens Contributes to Psychological Stress Resilience. *Molecular Psychiatry, 14*(5), 532–536. mp2008103 [pii]10.1038/mp.2008.103

McEwen, B. S. (1998). Protective and damaging effects of stress mediators. *The New England Journal of Medicine, 338*(3), 171–179.

Miller, G. E., Dopp, J. M., Myers, H. F., Stevens, S. Y., & Fahey, J. L. (1999). Psychosocial Predictors of Natural Killer Cell Mobilization During Marital Conflict. *Health Psychology, 18*(3), 262–271.

Moynihan, J. A., Larson, M. R., Treanor, J., Duberstein, P. R., Power, A., Shore, B., & Ader, R. (2004). Psychosocial Factors and the Response to Influenza Vaccination in Older Adults. *Psychosomatic Medicine, 66*(6), 950–953.

Munck, A., Guyre, P. M., & Holbrook, N. J. (1984). Physiological Functions of Glucocorticoids in Stress and Their Relation to Pharmacological Actions. *Endocrine Reviews, 5*(1), 25–44.

Ogata, K., An, E., Shioi, Y., Nakamura, K., Luo, S., Yokose, N., et al. (2001). Association Between Natural Killer Cell Activity and Infection in Immunologically Normal Elderly People. *Clinical and Experimental Immunology, 124*(3), 392–397.

Orentreich, N., Brind, J. L., Rizer, R. L., & Vogelman, J. H. (1984). Age-Changes and Sex-Differences in Serum Dehydroepiandrosterone Sulphate Concentrations Throughout Adulthood. *The Journal of Clinical Endocrinology and Metabolism, 59*(3), 551–555.

Orentreich, N., Brind, J. L., Vogelman, J. H., Andres, R., & Baldwin, H. (1992). Long-Term Longitudinal Measurements of Plasma Dehydroepiandrosterone Sulfate in Normal Men. *The Journal of Clinical Endocrinology and Metabolism, 75*(4), 1002–1004.

Pariante, C. M., Carpiniello, B., Orru, M. G., Sitzia, R., Piras, A., Farci, A. M. G., et al. (1997). Chronic Caregiving Stress Alters Peripheral Blood Immune Parameters: The Role of Age and Severity of Stress. *Psychotherapy and Psychosomatics, 66*(4), 199–207.

Phillips, A. C. (2011). Blunted Cardiovascular Reactivity Relates to Depression, Obesity, and Self-Reported Health. *Biological Psychology, 86*(2), 106–113. https://doi.org/10.1016/j.biopsycho.2010.03.016

Phillips, A. C., Burns, V. E., Carroll, D., Ring, C., & Drayson, M. (2005). The Association Between Life Events, Social Support and Antibody Status Following Thymus-Dependent and Thymus-Independent Vaccinations in Healthy Young Adults. *Brain, Behavior, and Immunity, 19*, 325–333.

Phillips, A. C., Carroll, D., Bums, V. E., Ring, C., Macleod, J., & Drayson, M. (2006a). Bereavement and Marriage Are Associated with Antibody Response to Influenza Vaccination in the Elderly. *Brain, Behavior, and Immunity, 20*(3), 279–289. https://doi.org/10.1016/j.bbi.2005.08.003

Phillips, A. C., Carroll, D., Evans, P., Bosch, J. A., Clow, A., Hucklebridge, F., & Der, G. (2006b). Stressful Life Events Are Associated with Low Secretion Rates of Immunoglobulin A in Saliva in the Middle Aged and Elderly. *Brain, Behavior, and Immunity, 20*, 191–197.

Phillips, A. C., Burns, V. E., & Lord, J. M. (2007). Stress and Exercise: Getting the Balance Right for an Ageing Immune System. *Exercise and Sport Sciences Reviews, 35*, 35–39.

Phillips, A. C., Upton, J., Duggal, N. A., Carroll, D., & Lord, J. M. (2013). Depression Following Hip Fracture Is Associated with Increased Physical Frailty in Older Adults: The Role of the Cortisol: Dehydroepiandrosterone Sulphate Ratio. *BMC Geriatrics, 13*(1), 60. doi:1471-2318-13-60 [pii]https://doi.org/10.1186/1471-2318-13-60

Pinquart, M., & Sorensen, S. (2003). Differences Between Caregivers and Noncaregivers in Psychological Health and Physical Health: A Meta-Analysis. *Psychology and Aging, 18*(2), 250–267. https://doi.org/10.1037/0882-7974.18.2.250

Pressman, S. D., Cohen, S., Miller, G. E., Barkin, A., Rabin, B. S., & Treanor, J. J. (2005). Loneliness, Social Network Size, and Immune Response to Influenza Vaccination in College Freshmen. *Health Psychology, 24*(3), 297–306.

Rasmussen, L. E. (1991). Gene Products of Cytomegalovirus and Their Immunologic Significance. In H. M (Ed.), *Cytomegalovirus: Biology and Infection* (2nd ed.). New York: Plenum Medical Book Company.

Robles, T. F., & Kiecolt-Glaser, J. K. (2003). The Physiology of Marriage: Pathways to Health. *Physiology & Behavior, 79*(3), 409–416.

Rosengren, A., Orth-Gomer, K., Wedel, H., & Wilhelmsen, L. (1993). Stressful Life Events, Social Support, and Mortality in Men Born in 1933. *British Medical Journal, 307*(6912), 1102–1105.

Sapolsky, R. M. (2007). Robert Sapolsky Discusses Physiological Effects of Stress. *Stanford News*. Retrieved from http://news.stanford.edu/news/2007/march7/sapolskysr-030707.html

Sapolsky, R. M., Romero, L. M., & Munck, A. U. (2000). How Do Glucocorticoids Influence Stress Responses? Integrating Permissive, Suppressive, Stimulatory, and Preparative Actions. *Endocrinological Review, 21*(1), 55–89.

Schleifer, S. J., Keller, S. E., Camerino, M., Thornton, J. C., & Stein, M. (1983). Suppression of Lymphocyte Stimulation Following Bereavement. *Jama-Journal of the American Medical Association, 250*(3), 374–377.

Sedova, L., Berube, J., Gaudet, D., Dumont, M., Tremblay, J., Hamet, P., & Pausova, Z. (2004). Diet-Induced Obesity Delays Cardiovascular Recovery from Stress in Spontaneously Hypertensive Rats. *Obesity Research, 12*(12), 1951–1958. https://doi.org/10.1038/oby.2004.245

Sheffield, D., McVey, C., & Carroll, D. (1996). Daily Events and Somatic Symptoms: Evidence of a Lagged Relationship. *The British Journal of Medical Psychology, 69*, 267–269.

Siergist, C. A. (2008). Vaccine Immunology. In S. Plotkin, W. Orenstein, & P. Offit (Eds.), *Vaccines* (pp. 17–36). Philadelphia: Saunders, Elsevier.

Sorrells, S. F., & Sapolsky, R. M. (2007). An Inflammatory Review of Glucocorticoid Actions in the CNS. *Brain, Behavior, and Immunity, 21*(3), 259–272. https://doi.org/10.1016/j.bbi.2006.11.006

Stahl, S. T., Arnold, A. M., Chen, J.-Y., Anderson, S., & Schulz, R. (2016). Mortality After Bereavement: The Role of Cardiovascular Disease and Depression. *Psychosomatic Medicine, 78*(6), 697–703. https://doi.org/10.1097/PSY.0000000000000317

Thompson, W. W., Shay, D. K., Weintraub, E., Brammer, L., Cox, N., Anderson, L. J., & Fukuda, K. (2003). Mortality Associated with Influenza and Respiratory Syncytial Virus in the United States. *Jama-Journal of the American Medical Association, 289*(2), 179–186. https://doi.org/10.1001/jama.289.2.179

Vedhara, K., Cox, N. K. M., Wilcock, G. K., Perks, P., Hunt, M., Anderson, S., et al. (1999a). Chronic Stress in Elderly Carers of Dementia Patients and

Antibody Response to Influenza Vaccination. *Lancet, 353*(9153), 627–631. https://doi.org/10.1016/s0140-6736(98)06098-x
Vedhara, K., Cox, N. K. M., Wilcock, G. K., Perks, P., Hunt, M., Anderson, S., et al. (1999b). Chronic Stress in Elderly Carers of Dementia Patients and Antibody Response to Influenza Vaccination. *Lancet, 20*, 627–631.
Vedhara, K., McDermott, M. P., Evans, T. G., Treanor, J. J., Plummer, S., Tallon, D., et al. (2002). Chronic Stress in Non-Elderly Caregivers: Psychological, Endocrine and Immune IMPLICATIONS. *Journal of Psychosomatic Research, 53*(6), 1153–1161.
Verbrugge, L. M. (1979). Marital Status and Health. *Journal of Marriage and the Family, 41*, 267–285.
Vitlic, A., Khanfer, R., Lord, J. M., Carroll, D., & Phillips, A. C. (2014). Bereavement Reduces Neutrophil Oxidative Burst Only in Older Adults: Role of the HPA Axis and Immunesenescence. *Immunity and Ageing, 11*(1).
Vitlic, A., Lord, J. M., Taylor, A. E., Arlt, W., Bartlett, D. B., Rossi, A., et al. (2015). Neutrophil Function in Young and Old Caregivers. *British Journal of Health Psychology, 21*, 173. https://doi.org/10.1111/bjhp.12156
von Kanel, R., Dimsdale, J. E., Mills, P. J., Ancoli-Israel, S., Patterson, T. L., Mausbach, B. T., & Grant, I. (2006). Effect of Alzheimer Caregiving Stress and Age on Frailty Markers Interleukin-6, C-Reactive Protein, and D-Dimer. *Journals of Gerontology Series a-Biological Sciences and Medical Sciences, 61*(9), 963–969.
Zisook, S., Shuchter, S. R., Irwin, M., Darko, D. F., Sledge, P., & Resovsky, K. (1994). Bereavement, Depression, and Immune Function. *Psychiatry Research, 52*(1), 1–10.

6

Healthy Lifestyles to Reduce Risk of Dementia

Jordan Elliott-King, Daniele Magistro, and Eef Hogervorst

Introduction

Biological psychology explores the biological bases of our experience and behaviour (Rosenzweig et al. 1996), while cognitive psychology investigates aspects of information processing. Together, both can contribute to our understanding of the psychology of ageing (Neisser 2014). Consistent growth in population ageing places importance on the biological and cognitive aspects of ageing, and through our understanding of these perspectives, it is possible to manage the risk of age-related diseases, such as dementia.

J. Elliott-King (✉) • E. Hogervorst
School of Sport, Exercise and Health Sciences, Loughborough University, Loughborough, UK
e-mail: j.elliott-king@lboro.ac.uk; E.Hogervorst@lboro.ac.uk

D. Magistro
Department of Sports Science, Nottingham Trent University, Nottingham, UK
e-mail: daniele.magistro@ntu.ac.uk

E. Peel et al. (eds.), *Psychologies of Ageing*,
https://doi.org/10.1007/978-3-319-97034-9_6

Brain changes are continuous throughout the lifespan (Pascual-Leone and Taylor 2011). Regions of the brain characterised by plasticity, or the ability to adapt and change, are more vulnerable to detrimental effects of normal ageing. This age-related vulnerability, however, renders selected brain regions susceptible to additional pathological changes that are not a part of normal ageing. It is a common misconception that dementia is a normal part of ageing (e.g., Glynn et al. 2017); dementia is in fact distinguishable with separate pathology. The most common form of dementia, Alzheimer's disease (AD) is characterised by amyloid plaques and tangles observable in the brain at post-mortem. Lewy Body dementia is characterised by the presence of Lewy bodies, and vascular dementia presents strokes and/or white matter disease. Overlap in pathologies at post-mortem is frequently observed (Hogervorst et al. 2003).

Dementia inhibits participation in activities of daily life and is typified by organic and acquired decline in cognitive functions, most notably memory impairments, and at least one other cognitive function without disturbance in consciousness (APA 2013). Dementia is the largest predictor of disability and dependence in older adults over the age of 65 (WHO 2012; Burns and Iliffe 2009) and prevalence is estimated at 6.4% for adults over 65 years of age (Lobo et al. 2000), which increases to 32% for those aged over 85 (Hebert et al. 2013). Currently, the only available medications are symptomatic drugs. However, these drugs are only effective for up to 30% of individuals and for a relatively short period of time, usually six months (Birks 2006; McShane et al. 2006). Targeting prevention through lifestyle behaviours is thus possibly the most effective approach to minimise the incidence of dementia.

Changes in brain structure and functioning caused by dementia pathology may begin years before clinical symptoms are observed. Preclinical brain changes engender mild changes in cognition that do not disrupt everyday life initially, but progress gradually. In the case of AD, mild cognitive impairment (MCI) may represent this open stage (Petersen et al. 1999). MCI refers to a clinical condition in which objective evidence of cognitive impairment in one or more cognitive domains is present, such as a slight decrease in memory, verbal skills, or orientation, but without a profound negative impact on daily activity, functioning, and personal independence (Gauthier et al. 2006; Petersen et al. 1999). Those with MCI are at higher risk of developing dementia; diligent management of MCI is therefore required to find strategies that produce effective prevention of dementia (Apostolo et al. 2016).

Normal ageing also involves cognitive changes due to neuron loss and alterations in synaptic density that increase with age (Dickstein et al. 2007; Hof and Morrison 2004). Many cognitive functions can be affected in a non-uniform manner during normal ageing, including attention and memory (Hedden and Gabrieli 2004). Certain aspects of memory are on average more affected by the ageing process. Episodic and working memory, for example, have been shown to be impaired in older adults (Kaschel et al. 2017; Mattay et al. 2006). Delayed free episodic recall is affected by the ageing process, while other memory functions, such as short-term memory storage (span) and semantic memory, remain relatively stable (Ishihara et al. 2002; Grady and Craik 2000).

Furthermore, older adults demonstrate attentional difficulties with selective and divided attention specifically (Commodari and Guarnera 2008). Selective attention is the ability to focus on a specific stimulus while inhibiting other non-salient stimuli. Everyday activities involve complex and multiple tasks simultaneously (Brustio et al. 2017), therefore selective attention, planning, execution of goal-directed behaviours, and judgement are important (Lezak 2004). The ability to divide attention between simultaneous tasks, which is affected by ageing, is also an important aspect of daily life (Shin and An 2014). Maintaining cognitive functioning is associated with the correct execution of daily physical tasks (Chodzko-Zajko et al. 2009; Tabbarah et al. 2002) and activities of daily living (ADL) (Candela et al. 2013).

Age-related cognitive decline does not occur equally between people and functions (Albert et al. 1995; Marden et al. 2017), with many older adults maintaining both good general and cognitive health (Kok et al. 2017; Rowe and Kahn 1997). Hence, cognitive decline cannot only be attributed to age. At any stage of life, including older age, actively engaging in healthy lifestyle choices can help to prevent cognitive decline, functional decline, extend longevity, enhance quality of life, and reduce risk for dementia (Forouzanfar et al. 2015). Although dementia is not a normal consequence of ageing, the risk factors for functional decline and dementia are shared. Dementia risk can be influenced by different factors including psychosocial factors, such as education, mental stimulation, social engagement, and physical activity. Additionally, vascular risk factors, such as high blood pressure, cholesterol, diabetes, and high body mass index have consistently been shown to increase the risk of late-onset dementia (WHO 2002).

Seven modifiable lifestyle factors have been discussed in the literature that explain approximately a third of all dementia cases worldwide. These are physical inactivity, midlife hypertension, midlife obesity, diabetes, low education, smoking, and depression (Norton et al. 2014). This chapter explores the most prominently discussed overarching lifestyle behaviours associated with dementia risk, which are physical activity, cognitive activity, and diet, as well as the feasibility of lifestyle changes in each of these categories and how challenges associated with ageing can present barriers to lifestyle change.

Physical Activity

Epidemiological studies show a lack of physical activity as important in the prediction of cognitive decline (Middleton et al. 2010) and dementia (Hamer and Chida 2009). Higher levels of physical activity show a relative reduction of dementia risk from 20% to 50% (Chodzko-Zajko et al. 2009; Kampert et al. 1996; Katzmarzyk et al. 2003). Different research methods provide varying levels of evidence. Longitudinal cohort studies, for instance, involve groups of people who share a defining characteristic taking part in assessments throughout their lives. Longitudinal cohort studies have shown that physical activity in individuals aged 55–65 predicts better performance on assessments of general cognition in later life (Almeida et al. 2006; Etgen et al. 2010). Studies have also shown better performance in episodic memory (Sabia et al. 2009; Stewart et al. 2003; Richards et al. 2003), processing speed (Chang et al. 2010; Stewart et al. 2003), and executive control (Chang et al. 2010; Sabia et al. 2009). Meta-analyses demonstrate a reduction of 38% in cognitive decline risk for healthy older adults with high levels of physical activity and 35% for older adults with low- to moderate levels of physical activity (Sofi et al. 2011), highlighting potential for preserving cognition through physical activity engagement (Prakash et al. 2015).

Longitudinal studies have further highlighted physical activity as effective in reducing cognitive decline in older people with MCI (Lytle et al. 2004; Scherder et al. 2005) and delaying dementia risk by three to six years (Abbott et al. 2004; Larson et al. 2006; Laurin et al. 2001). The majority of longitudinal studies with healthy older adults have found higher levels of physical activity to reduce risk of dementia by 30–50% (Barnes et al. 2007). The positive results observed may be magnified by participants engaging in multiple protective factors. Nonetheless, research so far substantiates the benefits of physical activity for cognitive health later in life.

Physical activity intervention studies with older adults have also found positive effects, including improved cognition (Angevaren et al. 2008; Barnes et al. 2007; Candela et al. 2015; Weuve et al. 2004; Yaffe et al. 2001), functional ability (Chodzko-Zajko et al. 2009; Magistro et al. 2014), and mental health (Penninx et al. 2000, 2002; Taylor et al. 2004). Sedentary older adults who completed an aerobic exercise programme for at least six months showed improvements in attention, memory, executive functioning, and speed of cognitive processing (Kramer et al. 2006; Colcombe and Kramer 2003; Candela et al. 2015), as well as general cognitive ability and speed of reasoning and memory (WHO 2002). A review of randomised controlled trials (RCTs) with healthy older adults has shown aerobic physical activity to be most beneficial for psychological wellbeing (Netz et al. 2005), memory functioning, cognitive speed, and auditory and visual attention (Angevaren et al. 2008). Memory and attention have been most consistently shown to improve with physical activity engagement, specifically long-term memory (Candela et al. 2015; Ruscheweyh et al. 2011) and selective attention (Owsley and McGwin 2004; Roth et al. 2003).

Numerous mechanisms have been proposed for how physical activity affects the brain. For older adults, physical activity has been shown to increase brain volume in both grey and white matter regions (Colcombe et al. 2006). Increased cardiovascular fitness, which can be achieved through aerobic activity, can improve brain plasticity and may serve to reduce both biological and cognitive ageing within the brain (Colcombe et al. 2004). There are further mechanisms that have been widely discussed, such as enhancement of neurotrophin levels, neurogenesis and vascularisation (Groot et al. 2016), among others, that could help to explain the brain's positive response to regular participation in physical activity. The frequency and intensity at which the activity is performed can also influence the outcome. Both high- and low-frequency physical activity show a positive effect on cognitive function (Groot et al. 2016), in a dose-dependent manner (Soni et al. 2017). Recommendations for physical activity derived from reviews promote moderate-intensity aerobic activity for older adults, muscle-strengthening activity, reducing sedentary behaviour, and managing risk of falls and injury (Nelson et al. 2007). Future research should seek to further develop our understanding of how the biological and cognitive aspects of physical activity can enhance cognitive functioning in the ageing brain and reduce dementia risk.

Cognitive Activity

The challenge of maintaining cognitive health has led to the emergence of more interventions designed to improve cognition. Evidence suggests that specific cognitive interventions for older adults increase neural plasticity (Johansson 2004; Pascual-Leone et al. 2005). This plasticity can then increase an individual's cognitive reserve (Stern 2012). Older adults who show high cognitive reserve, the ability to compensate and respond to brain degeneration and decline (Shah et al. 2017), demonstrate more resistance to dementia (Stern 2006, 2012) supported by a 50% reduction in dementia risk (Valenzuela and Sachdev 2006).

Observational studies show that high levels of cognitive activity could be beneficial in reducing cognitive decline and risk of dementia in older adults (Verghese et al. 2003). Longitudinal studies indicate that cognitive decline trajectories and outcomes could be mitigated by participating in cognitively stimulating activities (Hertzog et al. 2008; Hughes 2010). Daily engagement in cognitive activities can also predict preserved cognition (Ghisletta et al. 2006; Small et al. 2012) and decrease AD incidence (Wilson et al. 2002). Conjointly, decreases in engagement with cognitive activities over time have been shown to increase risk of cognitive decline (Mitchell et al. 2012).

RCT studies have highlighted cognitive activity as effective (Ball et al. 2002; Candela et al. 2015; Nouchi et al. 2012; Wolinsky et al. 2006).

Efficacy, however, is dependent on the task being trained. Cognitive training has improved targeted cognitive domains such as processing speed, memory, and reasoning in both healthy older adults and those reporting memory problems (Ball et al. 2002; Engvig et al. 2010; Mahncke et al. 2006; Nouchi et al. 2012). Several RCT studies also show cognitive training to be effective for memory and attention as these were the functions that were the focus of the training (Candela et al. 2015; Mozolic et al. 2011; Shah et al. 2017). Computerised cognitive training could also be beneficial for those with MCI or AD (Galante et al. 2007; Günther et al. 2003; Tárraga et al. 2006).

Usually, cognitive training, stimulation, and rehabilitative activity involve structured, repeated, and frequent engagement in standardised protocols of specific cognitive tasks, and are focused on specific cognitive domains. Domain-specific analyses underline the efficacy of the training for nonverbal memory, processing speed, working memory, and visuospatial outcomes, but not so much for attention and executive functions. However, efficacy is varied across the different cognitive domains, often due to intervention design (Lampit et al. 2014).

Meta-analyses discuss the importance of the activity setting, whether this is individually at home or in a group training set-up (Lampit et al. 2014). Home-based cognitive training is easier to administer, customise, and adapt, with potential for lower implementation cost (Kueider et al. 2012) and in many cases could facilitate the participation of frail and disabled older adults. Alternatively, group-based training that includes supervision by a trainer requires a suitable environment for delivery and is less personalised. However, in a group, service providers can ensure protocol compliance, adherence, provision of motivational support, encouragement, and social interaction. Overall, group activity could be more beneficial for older adults (Lampit et al. 2014; Verhaeghen et al. 1992).

Diet and Nutrition

Dietary choices and dementia risk have been increasingly studied. Recommendations suggest minimising intake of saturated and trans-fats (Barnard et al. 2014). Although one longitudinal cohort study found no association between risk of dementia and intake of total saturated fats, trans-fats, cholesterol, or alternative forms of fatty acids (Engelhart et al. 2002), other findings do suggest that intake of saturated fats increases the risk of developing dementia (e.g., Morris and Tangney 2014; Amadieu et al. 2017). This suggests that lower levels of saturated fat could reduce the risk for dementia, but evidence remains inconclusive. The contradictions highlighted here could stem from a number of different alternative forms of fatty acids that are suggested to protect against dementia. For instance, high intake of n-3 Polyunsaturated fatty acids (PUFA), may protect against age-related cognitive decline. A high intake of saturated fat and cholesterol, on the other hand, increases risk of cardiovascular disease, which secondarily increases dementia risk (Kalmijn et al. 1997).

In conjunction with fats, sugary foods and drinks such as sodas and fruit juices should be considered. Dementia researchers have used data from the Framingham heart study including 1484 adults aged 60 and over to assess longitudinally the relationship between sugar and dementia risk (Gordon et al. 1977). When examining cumulative beverage consumption, daily intake of artificially sweetened soft drinks was associated with an increased risk of dementia; however, such associations did not withstand after adjustment for covariates (Pase et al. 2017). Although this suggests a potential increase in risk through soda consumption, further investigation is warranted.

Evidence suggests that vegetables, legumes, fruits, and whole grains could incur positive health benefits including reducing dementia risk; recommendations have therefore advocated these foods replacing meats as primary staples of diet (Barnard et al. 2014). Not all vegetables, however, directly affect dementia-related mechanisms (Solfrizzi et al. 2010). Plant-based foods and leafy green vegetables, such as broccoli, kale, and spinach, as well as beans, peas, citrus fruits, cantaloupe, and melons, which are rich in vitamin B6, B12, and folate, are specifically noteworthy due to their role in the mechanisms involved in homocysteine, an amino acid found in the brain. Homocysteine in the brain can lead to detrimental effects if vitamin B6, B12, or folate are not available. Plasma homocysteine is a strong, independent risk factor for dementia (e.g., Seshadri et al. 2002). RCTs have found positive effects of folate and folic acid on cognition but only in people who were folate deficient (Lefevre-Arbogast et al. 2016; Hogervorst 2017).

Oxidative stress, an imbalance between oxidants and antioxidants in favour of the oxidants (Sies 1997), is another biological mechanism implicated in dementia risk (e.g., Bennett et al. 2009). Higher intake of vitamin E, an antioxidant, has been suggested to reduce oxidative stress and thus dementia risk (Barnard et al. 2014). Vitamin E occurs naturally and is found in many foods including mangoes, papaya, avocadoes, tomatoes, red bell peppers, spinach, nuts, seeds, and oils. However, evidence for the efficacy of vitamin E in dementia risk reduction is inconsistent (Crichton et al. 2013). There is sufficient evidence to indicate that dementia can be delayed by adequate intake of specific nutrients, but these could be most beneficial through whole foods rather than dietary supplements.

Moreover, supplements often contain iron and copper which, in excess, can contribute to cognitive problems (e.g., Squitti et al. 2014). Alongside the discussion of iron and copper is the ongoing debate regarding aluminium and dementia risk. Aluminium, found in pots, pans, and cooking equipment, has been weakly associated with dementia risk. Evidence is limited, but novel measurements of aluminium from the brain tissue of individuals with dementia show extremely high concentrations (Mirza et al. 2017). The mechanisms under which this association could occur are yet to be explained, but does substantiate advice to avoid aluminium intake to reduce dementia risk, which has only tentatively been issued in earlier guidelines (e.g., Barnard et al. 2014).

Recent literature has shifted to analyse dietary patterns as a whole, rather than specific nutrients, to provide a more holistic picture of dietary risk (e.g., Van de Rest et al. 2015). The most widely examined dietary pattern is the Mediterranean diet, essential components of which include monounsaturated fatty acids, derived from olive oil, vegetables, legumes, nuts, cheese, fruits, cereals and wine (Panza et al. 2004). Systematic review has associated greater adherence to the Mediterranean diet with slower cognitive decline and lower risk of developing AD (Lourida et al. 2013). Components of the Mediterranean and other suggested diets, however, do not contain all components that are specifically related to cognition. Reviews advocating the Mediterranean diet are based on improving overall health status or blood pressure, which elicit a secondary benefit to cognition rather than targeting cognition directly (Van de Rest et al. 2015). However, investigating whole-diet approaches is an attractive strategy due to the potential for combined nutrient effects to yield substantial outcomes.

When considering diet and dementia risk, whether the aim is to increase folate, B12, and B6; decrease saturated and trans-fat; and/or adhere to the whole Mediterranean diet, the literature has suggested that dietary interventions are best placed in midlife due to their potential to protect against developing dementia pathology (e.g., Van de Rest et al. 2015; Hogervorst 2017). Once people have developed MCI or dementia, nutritional interventions appear less successful (Hogervorst 2017).

Feasibility of Risk Reduction

Lifestyle changes are adaptations individuals make to their daily lives in order to incorporate more healthy behaviours such as engaging with more physical activity or eating more vegetables, for example. These changes render minimal risk of harm (Barnard et al. 2014), and although risk and protective factors can vary with age, the maintenance of a healthy lifestyle provides the best option for preventing dementia (Peters 2009). However, an individual's attitude concerning dementia risk may affect motivation to change health behaviours and lifestyle (Kim et al. 2014). Numerous barriers that can inhibit individuals from following lifestyle recommendations have been highlighted throughout the literature.

In a review, Kelly et al. (2016) established six themes that encompassed factors influencing uptake and maintenance of health behaviours. These were health and quality of life; sociocultural factors; the physical environment; access to facilities and resources, psychological factors, such as self-efficacy and motivation; and evidence relating to health inequalities. These overarching themes consist of specific barriers that can inhibit individuals from achieving a healthy lifestyle. For instance, a lack of time (this could be due to family, household, or occupational responsibilities); access to transport, facilities, or resources; financial costs;

entrenched attitudes and behaviours; restrictions that stem from the physical environment, being of a low socio-economic status, and a lack of knowledge about health are among the most commonly cited (Kelly et al. 2016). Additionally, specific demographic factors, such as age and family history, were shown throughout a multi-ethnic cohort to significantly determine an individual's willingness to engage with risk reduction (Seifan et al. 2017). Chapter 7 in this volume gives further background on factors that may prevent or facilitate the effectiveness of lifestyle behaviour change interventions.

Barriers that have been cited in the physical activity literature specifically include overall health, specific symptoms related to depression and pain, the general environment, neighbourhood crime rate, a lack of advice from physicians, knowledge, childhood experiences with physical activity, and marital hardships (Schutzer and Graves 2004; Schoeny et al. 2017). Similarly, specific barriers for diet and nutrition have also been discussed: time was cited as a key barrier to healthy eating, alongside concern about cost of more nutritious choices (Eikenberry and Smith 2004). Barriers could also be strengthened by strong intrinsic Western sociocultural values, norms, and traditions that could substantially impede on the feasibility of making lifestyle alterations (Knight et al. 2016).

Theories have sought to support initiation and maintenance of positive lifestyle behaviours as we age. Most recently, the upward spiral theory of lifestyle changes proposes that positive affect can facilitate long-term adherence to positive health behaviours through processes known as the inner and outer loop (Van Cappellen et al. 2018). The inner loop supposes that positive affect experienced when engaging with health behaviours increases the salience of the unconscious incentive to repeat that behaviour. The outer loop then further works to build up a suite of internal resources which amplify the positive affect experienced during positive health behaviours and therefore strengthen unconscious motives. Theories such as this aid explanation of positive affect experienced during positive health behaviours that encourage longer-term maintenance of healthy lifestyle behaviours. For this theory to take effect, health behaviours need to be engaged with in the first instance. Reasons that individuals might engage with lifestyle changes include being aware of the benefits of healthy ageing, a focus on the enjoyment, social and family support, clear messages, and integration of behaviours into pre-existing lifestyles (Kelly et al. 2016; Eikenberry and Smith 2004). See also Chap. 7 in this volume for discussion of other behaviour change theories.

Health education has been identified as an integral part of engagement with reducing risk for dementia. Teaching the benefits of healthy lifestyle behaviours has been shown to improve health and discourage behaviours that lead to poor health (Lafortune et al. 2017), as well as increase willingness to engage with lifestyle changes (Seifan et al. 2017). Review by Lafortune and colleagues revealed that tailoring physical activity interventions to include education has been found to incur substantial benefits to intervention success. A multimodal intervention, for instance, was shown to influence decisions to start and maintain physical activity, but was less effective than supervised exercise sessions. Furthermore, interventions that include support, self-monitoring, rewards, personal goal setting, and/or provision of information about local opportunities are more likely to be successful at instigating lifestyle change. The review discussed the use of printed advice as beneficial for dietary interventions (Lafortune et al. 2017). Moreover, supplementing interventions with between-session phone calls may be an effective way to strengthen intervention effects (Schoeny et al. 2017) and should therefore be applied where possible.

Throughout the literature, facilitating this reduction in dementia risk has often been discussed in parallel to visiting a healthcare professional (e.g., Seifan et al. 2017). Both men and women rated healthcare professionals as the most important source of support in changing lifestyle behaviours (Mosca et al. 1998), despite other sources of support being available. Healthcare professionals are suitably positioned to communicate lifestyle advice and act as a catalyst for health behaviour change. Although evidence regarding health behaviour change resulting from healthcare professional advice remains inconclusive (Eden et al. 2002), several studies have highlighted some success. For instance, Kreuter et al. (2000) found that individuals who received advice to quit smoking, eat less fat, or engage with more physical activity prior to receiving intervention materials on the same topic were far more likely to perceive the materials as pertinent to their health, remember the materials, and share them with others. Following this, they were also more likely to report making changes or trying to make changes to health behaviours.

Although risk factors for dementia are often termed modifiable, this is not always the case. The feasibility of lifestyle changes varies from one circumstance to the next. Therefore, careful consideration alongside an understanding of the barriers and facilitators to lifestyle change that might apply is required to ensure the intervention is well targeted and therefore has a greater chance of improving health and subsequently decreasing risk for dementia. Healthcare professionals can often play a role in providing advice and support to overcome barriers and are positioned to instigate lifestyle change for a number of individuals.

Conclusions

It is possible to reduce the risk of developing dementia through modifiable lifestyle factors. Factors discussed focused on physical activity, cognitive training, and diet. All factors showed benefits for healthy older adults and those with MCI, suggesting the potential for lifestyle factors to be beneficial in risk reduction regardless of current cognitive status.

Research has consistently demonstrated the importance of physical activity in dementia risk reduction, as well as overall improving health. With studies showing physical activity to reduce risk of dementia and resulting increases in brain volumes may incur additional health benefits of maintaining physical activity into old age (Tan et al. 2016). Among other factors, physical activity has been shown to positively influence quality of life, psychological wellbeing, and physical health, but, most crucially, it incurs benefits to overall cognition, as well as specific functions, such as attention, memory, executive functioning, speed of cognitive processing, and speed of reasoning and memory. Evidence so far thus substantiates the claim that physical activity is a powerful tool in preventing dementia, and that increases in physical activity should be sought throughout the lifespan. Most importantly, the majority of studies have shown that physical activity can improve cognition and independence in people with dementia, whereas for nutrition and cognitive training this is much less clear.

Growth of the global computerised brain health software market (Fernandez 2010) in recent years has highlighted the increase in the popularity of cognitive training. Although cognitive training research is still developing, evidence for increased neuroplasticity (Johansson 2004) resulting in potential increases in cognitive reserve (Stern 2012) suggests potential for cognitive training to play important role in dementia risk reduction (e.g., Verghese et al. 2003). Specific cognitive domains targeted by training have been shown to yield specific improvements, such as in nonverbal memory, processing speed, working memory, and visuospatial outcomes (Lampit et al. 2014). The literature has further suggested that cognitive training could be more suitably administered in a group, and practically challenging aspects, such as finding appropriate environments, time, and resources, could incur numerous benefits far beyond individual home-based training.

Lastly, the diet and nutrition literature has indicated that interventions would be most effectively positioned in midlife (e.g., Van de Rest et al. 2015). Saturated and trans-fats have been shown to have negative effects on cognition and intake and therefore should be minimised to reduce risk of developing dementia, although evidence regarding omega-3 is inconclusive. Vegetables, legumes, fruits, and whole grains have been found to benefit cognition and are suggested as an alternative to red meats as much as possible. A particular focus on consumption of plant-based foods that are high in B vitamins and folate was underlined due to the biological role these vitamins play in metabolising the neurotoxic homocysteine, but also by having direct protective effects on the brain and reducing risk of dementia. Studies examining the role of vitamin E, C, and A as antioxidants present inconsistent results. When consumed, however, for maximum benefit, whole foods should be sought rather than supplements, as the latter have often been shown to be ineffective. Discussions of whole dietary changes, rather than focus on individual nutrients, have increasingly been advocated throughout the literature due to the potential to combine effects of nutrients and yield more substantive outcomes. Positive indications stemming from the success of the Mediterranean diet have been highlighted thus far but further RCTs are required to establish whether cultural factors are influencing dietary benefits or if the diet alone is of significant benefit enough to warrant use as an intervention.

Considering the difficulties of implementing lifestyle changes is paramount to intervention success. Therefore, lifestyle and behaviour changes cannot be implemented without consideration for the feasibility of the changes, the barriers that could inhibit change, and the facilitators and motivators that could increase the likelihood of taking up and maintaining healthy lifestyle changes. The most consistent barriers emphasised for all lifestyle changes were health, time, costs, and environment. The most integral factor to help facilitate healthy lifestyle changes was education, with further intervention benefits shown with a focus on enjoyment and social support included in the intervention.

Implications for practice are substantial. The role of healthcare professionals in instigating lifestyle changes should be harnessed with direct advice pertaining to increases in physical activity and nutritional advice, particularly advocating increases in B vitamins, folate, and minimising saturated and trans-fat, throughout the lifespan. If interventions are adapted to aim to initiate and sustain healthy lifestyle choices, the risk of developing dementia can be drastically reduced and further health benefits could follow.

References

Abbott, R. D., White, L. R., Ross, G. W., Masaki, K. H., Curb, J. D., & Petrovitch, H. (2004). Walking and Dementia in Physically Capable Elderly Men. *JAMA, 292*(12), 1447–1453.

Albert, M. S., Jones, K., Savage, C. R., Berkman, L., Seeman, T., Blazer, D., & Rowe, J. W. (1995). Predictors of Cognitive Change in Older Persons: MacArthur Studies of Successful Aging. *Psychology and Aging, 10*(4), 578.

Almeida, O. P., Norman, P., Hankey, G., Jamrozik, K., & Flicker, L. (2006). Successful Mental Health Aging: Results from a Longitudinal Study of Older Australian Men. *The American Journal of Geriatric Psychiatry, 14*(1), 27–35.

Amadieu, C., Lefèvre-Arbogast, S., Delcourt, C., Dartigues, J. F., Helmer, C., Féart, C., & Samieri, C. (2017). Nutrient Biomarker Patterns and Long-Term Risk of Dementia in Older Adults. *Alzheimer's & Dementia: The Journal of the Alzheimer's Association, 13*(10), 1125–1132.

American Psychiatric Association. (2013). *Diagnostic and Statistical Manual of Mental Disorders (DSM-5®)*. Arlington: American Psychiatric Pub.

Angevaren, M., Aufdemkampe, G., Verhaar, H. J., Aleman, A., & Vanhees, L. (2008). Physical Activity and Enhanced Fitness to Improve Cognitive Function in Older People Without Known Cognitive Impairment. *Cochrane Database of Systematic Reviews, 3*(3), CD005381.

Apostolo, J., Holland, C., O'Connell, M. D., Feeney, J., Tabares-Seisdedos, R., Tadros, G., … Varela-Nieto, I. (2016). Mild Cognitive Decline. A Position Statement of the Cognitive Decline Group of the European Innovation Partnership for Active and Healthy Ageing (EIPAHA). *Maturitas, 83*, 83–93.

Ball, K., Berch, D. B., Helmers, K. F., Jobe, J. B., Leveck, M. D., Marsiske, M., … Unverzagt, F. W. (2002). Effects of Cognitive Training Interventions with Older Adults: A Randomized Controlled Trial. *JAMA, 288*(18), 2271–2281.

Barnard, N. D., Bush, A. I., Ceccarelli, A., Cooper, J., de Jager, C. A., Erickson, K. I., … Morris, M. C. (2014). Dietary and Lifestyle Guidelines for the Prevention of Alzheimer's Disease. *Neurobiology of Aging, 35*, S74–S78.

Barnes, D. E., Whitmer, R. A., & Yaffe, K. (2007). Physical Activity and Dementia: The Need for Prevention Trials. *Exercise and Sport Sciences Reviews, 35*(1), 24–29.

Bennett, S., Grant, M. M., & Aldred, S. (2009). Oxidative Stress in Vascular Dementia and Alzheimer's Disease: A Common Pathology. *Journal of Alzheimer's Disease, 17*(2), 245–257.

Birks, J. S. (2006). Cholinesterase Inhibitors for Alzheimer's Disease. *Cochrane Database of Systematic Reviews*, 1, Art. No.: CD005593.

Brustio, P. R., Magistro, D., Zecca, M., Liubicich, M. E., & Rabaglietti, E. (2017). Fear of Falling and Activities of Daily Living Function: Mediation Effect of Dual-Task Ability. *Aging & Mental Health, 22*(6), 1–6. https://doi.org/10.1080/13607863.2017.1318257

Burns, A., & Iliffe, S. (2009). Dementia. *British Medical Journal, 338*, b75.

Candela, F., Zucchetti, G., & Magistro, D. (2013). Individual Correlates of Autonomy in Activities of Daily Living of Institutionalized Elderly Individuals: An Exploratory Study in a Holistic Perspective. *Holistic Nursing Practice, 27*(5), 284–291.

Candela, F., Zucchetti, G., Magistro, D., & Rabaglietti, E. (2015). The Effects of a Physical Activity Program and a Cognitive Training Program on the Long-Term Memory and Selective Attention of Older Adults: A Comparative Study. *Activities, Adaptation, & Aging, 39*(1), 77–91.

Chang, M., Jonsson, P. V., Snaedal, J., Bjornsson, S., Saczynski, J. S., Aspelund, T., … Gudnason, V. (2010). The Effect of Midlife Physical Activity on Cognitive Function Among Older Adults: AGES—Reykjavik Study. *Journals of Gerontology Series A: Biomedical Sciences and Medical Sciences, 65*(12), 1369–1374.

Chodzko-Zajko, W. J., Proctor, D. N., Fiatarone Singh, M. A., Minson, C. T., Nigg, C. R., Salem, G. J., & Skinner, J. S. (2009). Exercise and Physical Activity for Older Adults. *Medicine & Science in Sports & Exercise, 41*(7), 1510–1530.

Colcombe, S., & Kramer, A. F. (2003). Fitness Effects on the Cognitive Function of Older Adults: A Meta-Analytic Study. *Psychological Science, 14*(2), 125–130.

Colcombe, S. J., Kramer, A. F., Erickson, K. I., Scalf, P., McAuley, E., Cohen, N. J., ... Elavsky, S. (2004). Cardiovascular Fitness, Cortical Plasticity, and Aging. *Proceedings of the National Academy of Sciences of the United States of America, 101*(9), 3316–3321.

Colcombe, S. J., Erickson, K. I., Scalf, P. E., Kim, J. S., Prakash, R., McAuley, E., ... Kramer, A. F. (2006). Aerobic Exercise Training Increases Brain Volume in Aging Humans. *The Journals of Gerontology Series A: Biological Sciences and Medical Sciences, 61*(11), 1166–1170.

Commodari, E., & Guarnera, M. (2008). Attention and Aging. *Aging Clinical and Experimental Research, 20*(6), 578–584.

Crichton, G. E., Bryan, J., & Murphy, K. J. (2013). Dietary Antioxidants, Cognitive Function and Dementia-a Systematic Review. *Plant Foods for Human Nutrition, 68*(3), 279–292.

Dickstein, D. L., Kabaso, D., Rocher, A. B., Luebke, J. I., Wearne, S. L., & Hof, P. R. (2007). Changes in the Structural Complexity of the Aged Brain. *Aging Cell, 6*(3), 275–284.

Eden, K. B., Orleans, T., Mulrow, C. D., Pender, N. J., & Teutsch, S. M. (2002). *Clinician Counseling to Promote Physical Activity.* Rockville: Agency for Healthcare Research and Quality.

Eikenberry, N., & Smith, C. (2004). Healthful Eating: Perceptions, Motivations, Barriers, and Promoters in Low-Income Minnesota Communities. *Journal of the American Dietetic Association, 104*(7), 1158–1161.

Engelhart, M. J., Geerlings, M. I., Ruitenberg, A., van Swieten, J. C., Hofman, A., Witteman, J. C., & Breteler, M. M. (2002). Dietary Intake of Antioxidants and Risk of Alzheimer Disease. *JAMA, 287*(24), 3223–3229.

Engvig, A., Fjell, A. M., Westlye, L. T., Moberget, T., Sundseth, Ø., Larsen, V. A., & Walhovd, K. B. (2010). Effects of Memory Training on Cortical Thickness in the Elderly. *NeuroImage, 52*(4), 1667–1676.

Etgen, T., Sander, D., Huntgeburth, U., Poppert, H., Förstl, H., & Bickel, H. (2010). Physical Activity and Incident Cognitive Impairment in Elderly Persons: The INVADE Study. *Archives of Internal Medicine, 170*(2), 186–193.

Fernandez, A. (2010). *Transforming Brain Health with Digital Tools to Assess, Enhance and Repair Cognition Across the Lifespan.* 2010 "State-of-the-Market" Report. San Francisco.

Forouzanfar, M. H., Alexander, L., Anderson, H. R., Bachman, V. F., Biryukov, S., Brauer, M., … Murray, C. J. (2015). Global, Regional, and National Comparative Risk Assessment of 79 Behavioural, Environmental and Occupational, and Metabolic Risks or Clusters of Risks in 188 Countries, 1990–2013: A Systematic Analysis for the Global Burden of Disease Study 2013. *The Lancet, 386*(10010), 2287–2323.

Galante, E., Venturini, G., & Fiaccadori, C. (2007). Computer-Based Cognitive Intervention for Dementia: Preliminary Results of a Randomized Clinical Trial. *Giornale Italiano di Medicina del Lavoro ed Ergonomia, 29*(3 Suppl B), B26–B32.

Gauthier, S., Reisberg, B., Zaudig, M., Petersen, R. C., Ritchie, K., Broich, K., … Cummings, J. L. (2006). Mild Cognitive Impairment. *The Lancet, 367*(9518), 1262–1270.

Ghisletta, P., Bickel, J.-F., & Lövdén, M. (2006). Does Activity Engagement Protect Against Cognitive Decline in Old Age? Methodological and Analytical Considerations. *The Journals of Gerontology Series B: Psychological Sciences and Social Sciences, 61*(5), 253–P261.

Glynn, R. W., Shelley, E., & Lawlor, B. A. (2017). Public Knowledge and Understanding of Dementia—Evidence from a National Survey in Ireland. *Age and Ageing, 46*(5), 865–869.

Gordon, T., Castelli, W. P., Hjortland, M. C., Kannel, W. B., & Dawber, T. R. (1977). High Density Lipoprotein as a Protective Factor Against Coronary Heart Disease: The Framingham Study. *The American Journal of Medicine, 62*(5), 707–714.

Grady, C. L., & Craik, F. I. (2000). Changes in Memory Processing with Age. *Current Opinion in Neurobiology, 10*(2), 224–231.

Groot, C., Hooghiemstra, A. M., Raijmakers, P. G. H. M., van Berckel, B. N. M., Scheltens, P., Scherder, E. J. A., … Ossenkoppele, R. (2016). The Effect of Physical Activity on Cognitive Function in Patients with Dementia: A Meta-Analysis of Randomized Control Trials. *Ageing Research Reviews, 25*, 13–23.

Günther, V. K., Schäfer, P., Holzner, B. J., & Kemmler, G. W. (2003). Long-term Improvements in Cognitive Performance Through Computer-Assisted Cognitive Training: A Pilot Study in a Residential Home for Older People. *Aging & Mental Health, 7*(3), 200–206.

Hamer, M., & Chida, Y. (2009). Physical Activity and Risk of Neurodegenerative Disease: A Systematic Review of Prospective Evidence. *Psychological Medicine, 39*(1), 3–11.
Hebert, L. E., Weuve, J., Scherr, P. A., & Evans, D. A. (2013). Alzheimer Disease in the United States (2010–2050) Estimated Using the 2010 Census. *Neurology, 80*(19), 1778–1783.
Hedden, T., & Gabrieli, J. D. E. (2004). Insights into the Ageing Mind: A View from Cognitive Neuroscience. *Nature Reviews Neuroscience, 5*(2), 87–96.
Hertzog, C., Kramer, A. F., Wilson, R. S., & Lindenberger, U. (2008). Enrichment Effects on Adult Cognitive Development: Can the Functional Capacity of Older Adults Be Preserved and Enhanced? *Psychological Science in the Public Interest, 9*(1), 1–65.
Hof, P. R., & Morrison, J. H. (2004). The Aging Brain: Morphomolecular Senescence of Cortical Circuits. *Trends in Neurosciences, 27*(10), 607–613.
Hogervorst, E. (2017). Healthy Lifestyles to Prevent Dementia and Reduce Dementia Symptoms. *Working with Older People, 21*(1), 31–39.
Hogervorst, E., Bandelow, S., Combrinck, M., Irani, S. R., & Smith, A. D. (2003). The Validity and Reliability of 6 Sets of Clinical Criteria to Classify Alzheimer's Disease and Vascular Dementia in Cases Confirmed Post-Mortem: Added Value of a Decision Tree Approach. *Dementia and Geriatric Cognitive Disorders, 16*(3), 170–180.
Hughes, T. F. (2010). Promotion of Cognitive Health Through Cognitive Activity in the Aging Population. *Aging Health, 6*(1), 111–121.
Ishihara, O., Gondo, Y., & Poon, L. W. (2002). The Influence of Aging on Short-Term and Long-Term Memory in the Continuous Recognition Paradigm. *Shinrigaku kenkyu: The Japanese Journal of Psychology, 72*(6), 516–521.
Johansson, B. B. (2004). Brain Plasticity in Health and Disease. *The Keio Journal of Medicine, 53*(4), 231–246.
Kalmijn, S., Launer, L. J., Ott, A., Witteman, J., Hofman, A., & Breteler, M. (1997). Dietary Fat Intake and the Risk of Incident Dementia in the Rotterdam Study. *Annals of Neurology, 42*(5), 776–782.
Kampert, J. B., Blair, S. N., Barlow, C. E., & Kohl, H. W. (1996). Physical Activity, Physical Fitness, and All-Cause and Cancer Mortality: A Prospective Study of Men and Women. *Annals of Epidemiology, 6*(5), 452–457.
Kaschel, R., Kazén, M., & Kuhl, J. (2017). State Orientation and Memory Load Impair Prospective Memory Performance in Older Compared to Younger Persons. *Aging, Neuropsychology, and Cognition, 24*(4), 453–469.

Katzmarzyk, P. T., Janssen, I., & Ardern, C. I. (*2003*). Physical Inactivity, Excess Adiposity and Premature Mortality. *Obesity Reviews, 4*(4), 257–290.

Kelly, S., Martin, S., Kuhn, I., Cowan, A., Brayne, C., & Lafortune, L. (2016). Barriers and Facilitators to the Uptake and Maintenance of Healthy Behaviours by People at Mid-Life: A Rapid Systematic Review. *PLoS One, 11*(1), e0145074.

Kim, S., Sargent-Cox, K., Cherbuin, N., & Anstey, K. J. (2014). Development of the Motivation to Change Lifestyle and Health Behaviours for Dementia Risk Reduction Scale. *Dementia and Geriatric Cognitive Disorders Extra, 4*(2), 172–183.

Knight, A., Bryan, J., & Murphy, K. (2016). Is the Mediterranean Diet a Feasible Approach to Preserving Cognitive Function and Reducing Risk of Dementia for Older Adults in Western Countries? New Insights and Future Directions. *Ageing Research Reviews, 25*, 85–101.

Kok, A. A., Aartsen, M. J., Deeg, D. J., & Huisman, M. (2017). Capturing the Diversity of Successful Aging: An Operational Definition Based on 16-Year Trajectories of Functioning. *The Gerontologist, 57*(2), 240–251.

Kramer, A. F., Erickson, K. I., & Colcombe, S. J. (2006). Exercise, Cognition, and the Aging Brain. *Journal of Applied Physiology, 101*(4), 1237–1242.

Kreuter, M. W., Chheda, S. G., & Bull, F. C. (2000). How Does Physician Advice Influence Patient Behavior?: Evidence for a Priming Effect. *Archives of Family Medicine, 9*(5), 426.

Kueider, A. M., Parisi, J. M., Gross, A. L., & Rebok, G. W. (2012). Computerized Cognitive Training with Older Adults: A Systematic Review. *PLoS One, 7*(*7*), e40588.

Lafortune, L., Martin, S., & Kelly, S. (2017). Midlife Approaches to Prevention of Ill Health in Later Life. *JAMA Internal Medicine, 177*(3), 423–424.

Lampit, A., Hallock, H., & Valenzuela, M. (2014). Computerized Cognitive Training in Cognitively Healthy Older Adults: A Systematic Review and Meta-Analysis of Effect Modifiers. *PLoS Medicine, 11*(*11*), e1001756.

Larson, E. B., Wang, L., Bowen, J. D., McCormick, W. C., Teri, L., Crane, P., & Kukull, W. (2006). Exercise Is Associated with Reduced Risk for Incident Dementia Among Persons 65 Years of Age and Older. *Annals of Internal Medicine, 144*(2), 73–81.

Laurin, D., Verreault, R., Lindsay, J., MacPherson, K., & Rockwood, K. (2001). Physical Activity and Risk of Cognitive Impairment and Dementia in Elderly Persons. *Archives of Neurology, 58*(3), 498–504.

Lefèvre-Arbogast, S., Féart, C., Dartigues, J. F., Helmer, C., Letenneur, L., & Samieri, C. (2016). Dietary B Vitamins and a 10-Year Risk of Dementia in Older Persons. *Nutrients, 8*(12), 761.

Lezak, M. D. (2004). *Neuropsychological Assessment*. Portland: Oxford University Press.

Lobo, A., Launer, L. J., Fratiglioni, L., Andersen, K., Di Carlo, A., Breteler, M. M. B., … Soininen, H. (2000). Prevalence of Dementia and Major Subtypes in Europe: A Collaborative Study of Population-Based Cohorts. *Neurology, 54*(5), S4.

Lourida, I., Soni, M., Thompson-Coon, J., Purandare, N., Lang, I. A., Ukoumunne, O. C., & Llewellyn, D. J. (2013). Mediterranean Diet, Cognitive Function, and Dementia: A Systematic Review. *Epidemiology, 24*(4), 479–489.

Lytle, M. E., Vander Bilt, J., Pandav, R. S., Dodge, H. H., & Ganguli, M. (2004). Exercise Level and Cognitive Decline: The Movies Project. *Alzheimer Disease & Associated Disorders, 18*(2), 57–64.

Magistro, D., Liubicich, M. E., Candela, F., & Ciairano, S. (2014). Effect of Ecological Walking Training in Sedentary Elderly People: Act on Aging Study. *The Gerontologist, 54*(4), 611–623.

Mahncke, H. W., Connor, B. B., Appelman, J., Ahsanuddin, O. N., Hardy, J. L., Wood, R. A., … Merzenich, M. M. (2006). Memory Enhancement in Healthy Older Adults Using a Brain Plasticity-Based Training Program: A Randomized, Controlled Study. *Proceedings of the National Academy of Sciences, 103*(33), 12523–12528.

Marden, J. R., Tchetgen Tchetgen, E. J., Kawachi, I., & Glymour, M. M. (2017). Contribution of Socioeconomic Status at 3 Life-Course Periods to Late-Life Memory Function and Decline: Early and Late Predictors of Dementia Risk. *American Journal of Epidemiology, 186*(7), 805–814.

Mattay, V. S., Fera, F., Tessitore, A., Hariri, A. R., Berman, K. F., Das, S., … Weinberger, D. R. (2006). Neurophysiological Correlates of Age-Related Changes in Working Memory Capacity. *Neuroscience Letters, 392*(1), 32–37.

McShane, R., Areosa Sastre, A., & Minakaran, N. (2006). Memantine for Dementia. *Cochrane Database of Systematic Reviews* (2). Art. No.: CD003154. doi: https://doi.org/10.1002/14651858.CD003154.pub5.

Middleton, L. E., Barnes, D. E., Lui, L.-Y., & Yaffe, K. (2010). Physical Activity Over the Life Course and Its Association with Cognitive Performance and Impairment in Old Age. *Journal of the American Geriatrics Society, 58*(7), 1322–1326.

Mirza, A., King, A., Troakes, C., & Exley, C. (2017). Aluminium in Brain Tissue in Familial Alzheimer's Disease. *Journal of Trace Elements in Medicine and Biology, 40*, 30–36.

Mitchell, M. B., Cimino, C. R., Benitez, A., Brown, C. L., Gibbons, L. E., Kennison, R. F., … Lindwall, M. (2012). Cognitively Stimulating Activities: Effects on Cognition Across Four Studies with up to 21 Years of Longitudinal Data. *Journal of Aging Research*, 2–12. https://doi.org/10.1155/2012/461592.

Morris, M. C., & Tangney, C. C. (2014). Dietary Fat Composition and Dementia Risk. *Neurobiology of Aging, 35*, S59–S64.

Mosca, L., McGillen, C., & Rubenfire, M. (1998). Gender Differences in Barriers to Lifestyle Change for Cardiovascular Disease Prevention. *Journal of Women's Health, 7*(6), 711–715.

Mozolic, J. L., Long, A. B., Morgan, A. R., Rawley-Payne, M., & Laurienti, P. J. (2011). A Cognitive Training Intervention Improves Modality-Specific Attention in a Randomized Controlled Trial of Healthy Older Adults. *Neurobiology of Aging, 32*(4), 655–668.

Neisser, U. (2014). *Cognitive Psychology: CLASSIC Edition*. Suffolk: Psychology Press.

Nelson, M. E., Rejeski, W. J., Blair, S. N., Duncan, P. W., Judge, J. O., King, A. C., … Castaneda-Sceppa, C. (2007). Physical Activity and Public Health in Older Adults: Recommendation from the American College of Sports Medicine and the American Heart Association. *Circulation, 116*(9), 1094.

Netz, Y., Wu, M. J., Becker, B. J., & Tenenbaum, G. (2005). Physical Activity and Psychological Well-Being in Advanced Age: A Meta-Analysis of Intervention Studies. *Psychology and Aging, 20*(2), 272.

Norton, S., Matthews, F. E., Barnes, D. E., Yaffe, K., & Brayne, C. (2014). Potential for Primary Prevention of Alzheimer's Disease: An Analysis of Population-Based Data. *The Lancet Neurology, 13*(8), 788–794.

Nouchi, R., Taki, Y., Takeuchi, H., Hashizume, H., Akitsuki, Y., Shigemune, Y., … Kawashima, R. (2012). Brain Training Game Improves Executive Functions and Processing Speed in the Elderly: A Randomized Controlled Trial. *PLoS One, 7*(1), e29676.

Owsley, C., & McGwin, G. (2004). Association Between Visual Attention and Mobility in Older Adults. *Journal of the American Geriatrics Society, 52*(11), 1901–1906.

Panza, F., Solfrizzi, V., Colacicco, A. M., D'introno, A., Capurso, C., Torres, F., … Capurso, A. (2004). Mediterranean Diet and Cognitive Decline. *Public Health Nutrition, 7*(7), 959–963.

Pascual-Leone, A., & Taylor, M. J. (2011). A Developmental Framework of Brain and Cognition from Infancy to Old Age. *Brain Topography, 24*, 183–186.

Pascual-Leone, A., Amedi, A., Fregni, F., & Merabet, L. B. (2005). The Plastic Human Brain Cortex. *Annual Review of Neuroscience, 28*, 377–401.

Pase, M. P., Himali, J. J., Beiser, A. S., Aparicio, H. J., Satizabal, C. L., Vasan, R. S., … Jacques, P. F. (2017). Sugar-and Artificially Sweetened Beverages and the Risks of Incident Stroke and Dementia. *Stroke*. https://doi.org/10.1161/STROKEAHA.116.016027.

Penninx, B. W., Deeg, D. J., van Eijk, J. T., Beekman, A. T., & Guralnik, J. M. (2000). Changes in Depression and Physical Decline in Older Adults: A Longitudinal Perspective. *Journal of Affective Disorders, 61*(1–2), 1–12.

Penninx, B. W. J. H., Rejeski, W. J., Pandya, J., Miller, M. E., Di Bari, M., Applegate, W. B., & Pahor, M. (2002). Exercise and Depressive Symptoms a Comparison of Aerobic and Resistance Exercise Effects on Emotional and Physical Function in Older Persons with High and Low Depressive Symptomatology. *The Journals of Gerontology Series B: Psychological Sciences and Social Sciences, 57*(2), P124–P132.

Peters, R. (2009). The Prevention of Dementia. *International Journal of Geriatric Psychiatry, 24*(5), 452–458.

Petersen, R. C., Smith, G. E., Waring, S. C., Ivnik, R. J., Tangalos, E. G., & Kokmen, E. (1999). Mild Cognitive Impairment: Clinical Characterization and Outcome. *Archives of Neurology, 56*(3), 303–308.

Prakash, R. S., Voss, M. W., Erickson, K. I., & Kramer, A. F. (2015). Physical Activity and Cognitive Vitality. *Annual Review of Psychology, 66*, 769–797.

Richards, M., Hardy, R., & Wadsworth, M. E. (2003). Does Active Leisure Protect Cognition? Evidence from a National Birth Cohort. *Social Science & Medicine, 56*(4), 785–792.

Rosenzweig, M. R., Leiman, A. L., & Breedlove, S. M. (1996). *Psychobiology of Mental Disorders. Biological Psychology* (pp. 581–587). Sunderland: Sinauer Associates.

Roth, D. L., Goode, K. T., Clay, O. J., & Ball, K. K. (2003). Association of Physical Activity and Visual Attention in Older Adults. *Journal of Aging and Health, 15*(3), 534–547.

Rowe, J. W., & Kahn, R. L. (1997). Successful Aging. *The Gerontologist, 37*(4), 433–440.

Ruscheweyh, R., Willemer, C., Krüger, K., Duning, T., Warnecke, T., Sommer, J., … Flöel, A. (2011). Physical Activity and Memory Functions: An Interventional Study. *Neurobiology of Aging, 32*(7), 1304–1319.

Sabia, S., Nabi, H., Kivimaki, M., Shipley, M. J., Marmot, M. G., & Singh-Manoux, A. (2009). Health Behaviors from Early to Late Midlife as Predictors

of Cognitive Function: The Whitehall II Study. *American Journal of Epidemiology, 170*(4), 428–437.
Scherder, E. J., Van Paasschen, J., Deijen, J.-B., Van Der Knokke, S., Orlebeke, J. F. K., Burgers, I., … Sergeant, J. A. (2005). Physical Activity and Executive Functions in the Elderly with Mild Cognitive Impairment. *Aging & Mental Health, 9*(3), 272–280.
Schoeny, M. E., Fogg, L., Buchholz, S. W., Miller, A., & Wilbur, J. (2017). Barriers to Physical Activity as Moderators of Intervention Effects. *Preventive Medicine Reports, 5*, 57–64.
Schutzer, K. A., & Graves, B. S. (2004). Barriers and Motivations to Exercise in Older Adults. *Preventive Medicine, 39*(5), 1056–1061.
Seifan, A., Ganzer, C. A., Vermeylen, F., Parry, S., Zhu, J., Lyons, A., … Kim, S. (2017). Development and Validation of the Alzheimer's Prevention Beliefs Measure in a Multi-Ethnic Cohort—A Behavioral Theory Approach. *Journal of Public Health, 39*(4), 863–873.
Seshadri, S., Beiser, A., Selhub, J., Jacques, P. F., Rosenberg, I. H., D'agostino, R. B., … Wolf, P. A. (2002). Plasma Homocysteine as a Risk Factor for Dementia and Alzheimer's Disease. *New England Journal of Medicine, 346*(7), 476–483.
Shah, T. M., Weinborn, M., Verdile, G., Sohrabi, H. R., & Martins, R. N. (2017). Enhancing Cognitive Functioning in Healthy Older Adults: A Systematic Review of the Clinical Significance of Commercially Available Computerized Cognitive Training in Preventing Cognitive Decline. *Neuropsychology Review, 27*(1), 62–80.
Shin, S.-S., & An, D.-H. (2014). The Effect of Motor Dual-Task Balance Training on Balance and Gait of Elderly Women. *Journal of Physical Therapy Science, 26*(3), 359–361.
Sies, H. (1997). Oxidative Stress: Oxidants and Antioxidants. *Experimental Physiology, 82*(2), 291–295.
Small, B. J., Dixon, R. A., McArdle, J. J., & Grimm, K. J. (2012). Do Changes in Lifestyle Engagement Moderate Cognitive Decline in Normal Aging? Evidence from the Victoria Longitudinal Study. *Neuropsychology, 26*(2), 144.
Sofi, F., Valecchi, D., Bacci, D., Abbate, R., Gensini, G. F., Casini, A., & Macchi, C. (2011). Physical Activity and Risk of Cognitive Decline: A Meta-Analysis of Prospective Studies. *Journal of Internal Medicine, 269*(1), 107–117.
Solfrizzi, V., Frisardi, V., Capurso, C., D'Introno, A., Colacicco, A. M., Vendemiale, G., … Panza, F. (2010). Dietary Fatty Acids in Dementia and Predementia Syndromes: Epidemiological Evidence and Possible Underlying Mechanisms. *Ageing Research Reviews, 9*(2), 184–199.

Soni, M., Orrell, M., Bandelow, S., Steptoe, A., Rafnsson, S., d'Orsi, E., … Hogervorst, E. (2017). Physical Activity Pre-and Post-Dementia: English Longitudinal Study of Ageing. *Aging & Mental Health*. doi: https://doi.org/10.1080/13607863.2017.1390731

Squitti, R., Siotto, M., & Polimanti, R. (2014). Low-Copper Diet as a Preventive Strategy for Alzheimer's Disease. *Neurobiology of Aging, 35*, S40–S50.

Stern, Y. (2006). Cognitive Reserve and Alzheimer Disease. *Alzheimer Disease & Associated Disorders, 20*(2), 112–117.

Stern, Y. (2012). Cognitive Reserve in Ageing and Alzheimer's Disease. *The Lancet Neurology, 11*(11), 1006–1012.

Stewart, R., Prince, M., & Mann, A. (2003). Age, Vascular Risk, and Cognitive Decline in an Older, British, African-Caribbean Population. *Journal of the American Geriatrics Society, 51*(11), 1547–1553.

Tabbarah, M., Crimmins, E. M., & Seeman, T. E. (2002). The Relationship Between Cognitive and Physical Performance MacArthur Studies of Successful Aging. *The Journals of Gerontology Series A: Biological Sciences and Medical Sciences, 57*(4), M228–M235.

Tan, Z. S., Spartano, N. L., Beiser, A. S., DeCarli, C., Auerbach, S. H., Vasan, R. S., & Seshadri, S. (2016). Physical Activity, Brain Volume, and Dementia Risk: The Framingham Study. *Journals of Gerontology Series A: Biomedical Sciences and Medical Sciences, 72*(6), 789–795.

Tárraga, L., Boada, M., Modinos, G., Espinosa, A., Diego, S., Morera, A., … Becker, J. T. (2006). A Randomised Pilot Study to Assess the Efficacy of an Interactive, Multimedia tool of Cognitive Stimulation in Alzheimer's Disease. *Journal of Neurology, Neurosurgery & Psychiatry, 77*(10), 1116–1121.

Taylor, A. H., Cable, N. T., Faulkner, G., Hillsdon, M., Narici, M., & Van Der Bij, A. K. (2004). Physical Activity and Older Adults: A Review of Health Benefits and the Effectiveness of Interventions. *Journal of Sports Sciences, 22*(8), 703–725.

Valenzuela, M. J., & Sachdev, P. (2006). Brain Reserve and Dementia: A Systematic Review. *Psychological Medicine, 36*(4), 441–454.

Van Cappellen, P., Rice, E. L., Catalino, L. I., & Fredrickson, B. L. (2018). Positive Affective Processes Underlie Positive Health Behaviour Change. *Psychology & Health, 33*(1), 77–97.

Van de Rest, O., Berendsen, A. A., Haveman-Nies, A., & de Groot, L. C. (2015). Dietary Patterns, Cognitive Decline, and Dementia: A Systematic Review. *Advances in Nutrition: An International Review Journal, 6*(2), 154–168.

Verghese, J., Lipton, R. B., Katz, M. J., Hall, C. B., Derby, C. A., Kuslansky, G., … Buschke, H. (2003). Leisure Activities and the Risk of Dementia in the Elderly. *New England Journal of Medicine, 348*(25), 2508–2516.

Verhaeghen, P., Marcoen, A., & Goossens, L. (1992). Improving Memory Performance in the Aged Through Mnemonic Training: A Meta-Analytic Study. *Psychology and Aging, 7*(2), 242.

Weuve, J., Kang, J. H., Manson, J. E., Breteler, M. M., Ware, J. H., & Grodstein, F. (2004). Physical Activity, Including Walking, and Cognitive Function in Older Women. *JAMA, 292*(12), 1454–1461.

Wilson, R. S., De Leon, C. F. M., Barnes, L. L., Schneider, J. A., Bienias, J. L., Evans, D. A., & Bennett, D. A. (2002). Participation in Cognitively Stimulating Activities and Risk of Incident Alzheimer Disease. *JAMA, 287*(6), 742–748.

Wolinsky, F. D., Unverzagt, F. W., Smith, D. M., Jones, R., Stoddard, A., & Tennstedt, S. L. (2006). The ACTIVE Cognitive Training Trial and Health-Related Quality of Life: Protection that Lasts for 5 Years. *The Journals of Gerontology Series A: Biological Sciences and Medical Sciences, 61*(12), 1324–1329.

World Health Organization. (2002). *Active Ageing: A Policy Framework.* Geneva: World Health Organization.

World Health Organization. (2012). *Dementia: A Public Health Priority.* Geneva: World Health Organization. pp.viii + 102 pp. ref.384.

Yaffe, K., Barnes, D., Nevitt, M., Lui, L.-Y., & Covinsky, K. (2001). A Prospective Study of Physical Activity and Cognitive Decline in Elderly Women: Women Who Walk. *Archives of Internal Medicine, 161*(14), 1703–1708.

7

Frailty and Resilience: Are They Necessarily Mutually Exclusive?

Carol Holland, Ian Garner, and Holly Gwyther

Introduction

This chapter first considers the common clinical definitions and associated epidemiology of frailty. The crucial issue of frailty as potentially preventable or reversible is then discussed, using qualitative evidence to understand the perceptions of those involved with frailty, including frail and robust older people, formal and informal carers, healthcare practitioners, and policymakers. The chapter then introduces the concepts of

C. Holland (✉)
School of Health and Medicine, Lancaster University, Lancaster, UK
e-mail: c.a.holland@lancaster.ac.uk

I. Garner
School of Life and Health Sciences, Aston University, Birmingham, UK
e-mail: garneri@aston.ac.uk

H. Gwyther
Centre for Ageing Research, Division of Health Research,
Lancaster University, Lancaster, UK
e-mail: h.gwyther@lancaster.ac.uk

E. Peel et al. (eds.), *Psychologies of Ageing*,
https://doi.org/10.1007/978-3-319-97034-9_7

health behaviour change and the role of health psychology theory in the design of interventions for frailty. Examples of both quantitative and qualitative research, along with their synthesis, are used to illustrate how health psychology may use multiple methods to provide recommendations for what works for whom and in what circumstances. The variation in outcomes for people diagnosed with physical frailty is then considered in terms of examining the role of psychological frailty and the common co-existence of frailty with depression and with cognitive impairment. Returning to qualitative evidence, we then reconsider the concept of frailty as the opposite of resilience, for example, by asking how some older people maintain resilience and well-being in the face of significant physical impairment and if it is possible to not only intervene with frailty itself but also to support resilience.

Background: What Is Frailty?

Frailty is a recognised syndrome characterised as a state of increased vulnerability to adverse health outcomes when exposed to a stressor, for example, a new chronic disease diagnosis, an acute infection, or a fall (Clegg et al. 2013) or other types of trauma. For instance, health status is an important predictor for mortality following a significant bereavement (Bowling 2009). Frailty is a dynamic process as opposed to a binary diagnosis of frail or not frail, describing a transition from robustness through a pre-frail condition (which is often not identified) to a frail condition (Lang et al. 2009). As a person becomes frail, physiological reserves reduce and individuals become less active or able to cope both with basic independence activities of daily living and also any acute illnesses or physiological stressors. For example, if a frail older person catches influenza, they are much more likely to die from that illness than a person of the same age who is not frail. Therefore, frailty is conceptualised as an absence of resilience, and the two are often seen as residing at either end of a state of the balance that is health, with Morley (2013) defining frailty as an imbalance between stressors and resilience reserves.

Clinical definitions of frailty are still debated, but the two most widely accepted and cited approaches are: a physical phenotype, based largely on muscle weakness and loss of muscle mass (sarcopenia) (Fried et al. 2001; Fried et al. 2004), and an approach that focuses on the impact of an accumulation of deficits with a fuller profile of contributors to frailty but still including the physical weakness of the phenotype (Rockwood and Mitnitski 2007, 2011). These approaches, the frailty phenotype and the frailty profile respectively, have largely been validated in terms of how well they predict the adverse outcomes of death, institution admittance, and hospitalisation, leading to a conceptualisation of frailty progression as a "one way street" and as a marker for the proximity of the end of life. This medicalised approach does not take into account the contribution of a person's environment, personality, or social situation in their ability to be resilient. As this field has developed, some researchers have widened the concept of frailty to include psychological components such as cognitive function and mental health, or social factors such as lack of social contact or support, and perceived well-being, recognising that it is not a purely biological syndrome (e.g., Rodríguez-Mañas et al. 2013). Research has specifically highlighted the roles of psychological and social variables in frailty development (Levers et al. 2006; Kele, 2012; Young et al. 2016), with the focus on physical and cognitive factors in frailty assessment possibly because they are tangible, can be objectively identified, and are potentially treatable through medical means (Lally and Crome 2007).

Clinical Definitions and Diagnosis In detail, Fried's phenotype approach conceptualises frailty as a biological syndrome resulting from deficits in five physiological domains: global weakness (assessed by measures such as grip strength or time to raise oneself from a seated position), overall slowness (e.g., gait speed), exhaustion (self-reported), low physical activity (self-reported), and unintentional weight loss. Having two of these symptoms indicates a pre-frail or at-risk status and three or more of these symptoms indicate a frail state. Alternatively, a frailty profile, or index, (Rockwood and Mitnitski 2007) is a continuous, as opposed to a categorical, variable which is constructed by calculating the proportion of a set of health deficits and diagnoses manifest in the individual. Deficits are selected to be associated with increasing age but not age saturated—for example, a diagnosis of type 2 diabetes would be appropriate but normal age-related vision changes such as presbyopia would not. Importantly, this model incorporates issues such as polypharmacy, cognitive impairments, and geriatric syndromes (e.g., falls, delirium, and urinary incontinence) in addition to physical weakness and impairments.

Estimates of frailty prevalence in older adults (over 65 years) vary from 4% to 17%, with the prevalence of pre-frailty ranging from 19% to 59% (Collard et al. 2012). These two main approaches identify an overlapping population at high risk of poor prognosis in terms of outcomes such as hospital admissions, admission to nursing homes, and death. Frailty is also associated with high demand for formal and informal healthcare services (Young 2003), social and community services, and other resources (Auyeung et al. 2011; Fried et al. 2001; Rockwood et al. 2006). Finally, it is also associated with reduced quality of life (Holland et al. 2015) and depression, as detailed later in this chapter.

Frailty Is Malleable

However, frailty is dynamic and a process, rather than a fixed diagnosis. Increasingly, research suggests that it is a malleable and manageable condition, and that it may be possible to prevent, halt, or even reverse its progress and to manage or prevent its adverse consequences (Cameron et al. 2013; Ferrucci et al. 2004; Gill et al. 2002; Ng et al. 2015; Theou et al. 2011). Consequently, many interventions have been trialled. Physical exercise, nutrition interventions, combined exercise and nutrition, and geriatric evaluation and management strategies (GEM) which include personalised care pathways and multidisciplinary teams have all been shown to have some impact on measures of frailty (for a review, see Apostolo et al. 2017). Of these, muscle strengthening exercise, particularly resistance training, has been found to be beneficial in terms of preventing and treating frailty and improving functional performance (Cadore et al. 2014; Chan et al. 2012; Ng et al. 2015). A Cochrane review that specifically examined progressive resistance training found reliable effects (Lui and Latham 2009). In a significant longitudinal study on physically active lifestyle with five years follow-up, Peterson et al. (2009) evaluated the influence of different types and degrees of physically active lifestyles on the development/worsening of frailty, distinguishing between an active lifestyle, actual purposeful exercise involvement (e.g., taking part in exercise classes), and a sedentary lifestyle. For those who were not frail at the beginning of the study, people in both active lifestyle and exercise groups were less likely to become frail at the later time points, but amongst those who were already frail, only the purposeful exercise prevented worsening of frailty.

In contrast, changes to diet or supplementation alone have not been shown to be as successful in preventing or reversing frailty without an associated exercise component. A review by Zuliani et al. (2015) concluded that exercise (they indicated aerobic exercise as well as resistance training) in combination with an adequate protein and energy intake was needed to prevent and manage sarcopenia. The review by Apostolo et al. (2017) also concluded that nutritional supplementation as an intervention for frailty was only reliable when it was in combination with exercise but also where there was a nutritional deficiency in the first place, notably with vitamin D supplementation.

In a survey of frailty interventions and projects registered in the European Innovation Partnership on Active and Healthy Ageing (EIP-AHA) repository of Good Practices, Gwyther et al. (2017) found that the majority of projects focus on pre-frail or robust older people in terms of prevention. This may suggest a possible understanding of a critical clinical window in which to have an impact (Topinková 2008) but also suggests that many clinicians and researchers may not feel that intervention in significantly frail people is worthwhile. However, our recent review of interventions (Apostolo et al. 2017) did not find any sense of limit beyond which intervention was not effective. For example, one study (Cadore et al. 2014) found a significant impact of an exercise programme on components of frailty such as walking speed and grip strength for frail older people living in long-term care with an average age of 92 years.

A Role for Health Psychology in Improving Acceptability and Effectiveness of Frailty Intervention

An issue identified in the searches for the systematic review (Apostolo et al. 2017) was that very large numbers of interventions were found in the search which were interventions for frail older people but were not interventions for frailty (i.e., measures of frailty were not part of any assessed outcomes). A reason for this, and for the preponderance of interventions at earlier stages found in reviews and surveys (Gwyther et al. 2017), may be that frailty screening or intervention may not be seen as beneficial, as economically viable, or even acceptable to older adults, to their caregivers, to health and social care staff, or even to health policymakers. This may stem from a lack of belief in their effectiveness, or necessity, or from perceived barriers to involvement on the part of older people and their carers. It may also stem from barriers to time investment or funding the initiatives on the part of health and social care practitioners and policy makers. Older adults are more likely to refuse healthcare services than younger adults (Littlechild and Glasby 2000), and research has indicated a commonly found list of barriers to participation in healthy lifestyle activities or interventions, often related to cost, accessibility, and transport difficulties for older adults (Holland et al. 2008; Waller et al. 2012), but also to a reluctance to seek help even where it is acknowledged it is required, related to fear of stereotyping or loss of perceived independence (Shaw et al. 2016a).

Health psychology contributions to evaluations of the effectiveness of screening and interventions in other health areas have indicated issues that can have an impact on the success of the interventions, shifting the focus of the evaluation not just to measured outcomes but also to the implementation of the interventions (e.g., Oakley et al. 2006; Craig et al. 2008). These issues include belief in the efficacy of the intervention (UK CRC 2003) and expectations about its implementation, structure, or deliverability, and the self-efficacy of the professional or patient. For example, Bleijenberg et al. (2013) explored the expectations and experiences of practice nurses and general practitioners (GPs) in providing proactive care for frail older patients. Some of the practice nurses in this study initially reported a reduction in work satisfaction with a new care programme because it made them feel "insecure" while barriers were perceived as being a lack of time and financial compensation for the GPs. In another study, nurses initially suggested a barrier was lack of time but then clarified by suggesting a lack of confidence in their ability or mindset to provide the psychological support their patients and their relatives needed: "Once in a while it is lack of time. Though I think it sounds a bit crude, too, to say that we haven't got time, but sometimes it is [...] no, not time: we are not geared to talk to them in that way" (Lindhardt et al. 2008, p. 674).

These issues can all have an impact on the way in which the healthcare practitioner may deliver the screening or intervention (fidelity to protocol) and the adherence and compliance of the patient or participant to the intervention (Holland et al. 2013; Shaw et al. 2016b).

Research Example 1: Qualitative Studies of Beliefs and Experiences

For these reasons, a series of studies that formed part of the FOCUS project, funded by the EU commission,[1] set out to understand approaches to and beliefs around frailty screening and interventions amongst all stakeholders, ranging across frail and healthy older adults, formal and informal carers, healthcare practitioners and high-level healthcare policy-makers. The aim was to clarify the issues faced by professionals seeking to implement frailty interventions or by older adults seeking to reduce or prevent frailty. This was in order that the feasibility and effectiveness of interventions and prevention could be improved, viewed positively, and healthy ageing supported. Increasingly within the broader study of ageing, significant challenges are being addressed by multidisciplinary teams. In this group of multiple method studies, health psychologists and psychology of ageing experts, medical practitioners, and nursing practitioners/researchers worked together to ensure all perspectives were included and to provide the best environment for impact across the fields. Three of the FOCUS studies are considered here:

(1) a meta-synthesis of qualitative studies on stakeholders' views and experiences of care and interventions addressing frailty and pre-frailty

[1] Acknowledgments: Holly Gwyther is supported by the FOCUS project: *Frailty Management Optimisation through EIP-AHA Commitments and Utilisation of Stakeholders Input (FOCUS)*, funded by the EU (Grant number 664367 FOCUS) (http://focus-aha.eu/en/home). The authors are grateful to the full FOCUS team for their part in the cited publications that form some of the background to this chapter. Ian Garner is supported by a 50th Anniversary PhD studentship jointly funded by Aston University and the ExtraCare Charitable Trust.

(where stakeholders were frail and healthy older adults, formal and informal carers, and healthcare practitioners) (D'Avanzo et al. 2017);
(2) an inductive thematic analysis conducted on the focus group and individual interviews in three EU countries (Italy, Poland, UK) with five groups of stakeholders—frail and non-frail older adults, family caregivers, and health and social care professionals (Shaw et al. 2017); and
(3) an inductive thematic analysis conducted on semi-structured interviews with senior healthcare policy makers based in six European Countries (Belgium, Luxembourg, Italy, Spain, Poland, UK) (Gwyther et al. 2018).

These three studies each aimed to achieve learning that could feed into the optimisation of frailty prevention and treatment by understanding stakeholders' experiences, beliefs, and attitudes. This qualitative approach is an important one in combination with the quantitative methods employed to determine the actual effectiveness of interventions (see Apostolo et al. 2017 and Zuliani et al. 2015 and examples in Chap. 6). It adds to such information by exploring the reasons behind intervention effectiveness that are to do with stakeholders' personal experiences, perceptions, understandings, and responses or intentions, rather than their underlying biological, cognitive, or social healthcare processes. This approach is more person-centred, giving participants an opportunity to explain their responses to their own health changes and their own understandings of what health and frailty mean from their own personal, cultural, and social perspectives.

The meta-synthesis of 28 qualitative studies identified four themes within the stakeholders' experiences: *Uncertainty about malleability of frailty*; *Strategies to prevent or to respond to frailty*; *The capacity to care and family-centred service provision*; and *Power and choice* (D'Avanzo et al. 2017). The first theme, uncertainty about the malleability of frailty, was clear across the range of stakeholders. For example, although most healthcare practitioners recognised that frailty can be prevented and treated, they expressed uncertainty about the optimal timing of interventions or who is most likely to benefit. Notably, some of this uncertainty stemmed from a belief that after a certain point, frailty was no longer malleable:

"We're always dealing with the consequences, by the time we get to it, it is too late" (physiotherapist; Roland et al. 2011, p. 11).

Healthy older adults themselves thought of frailty as both inevitable and at least partially a consequence of long-term lifestyle, the effects of which, they felt, may not be possible to undo; "Frailty is not something that you can prevent, you cannot do anything, it just happens when you get older" (frail respondent; Puts, Shekary et al. 2009, p. 264). Frailer older adults and their carers were also dubious about the benefits of interventions: "They can't take my fear of falling away" (frail older adult; Faes et al. 2010, p. 839); "She is already 80 years old. With all her medical problems, such a programme would be useless" (caregiver; Faes et al. 2010, p. 839).

These findings were mirrored in the second study of this series, the focus group study on stakeholders' experiences of frailty and the meanings of frailty to individuals (Shaw et al. 2017). Doubt about the malleability of frailty was again clear in the discussions with several of the stakeholder groups, with one healthcare practitioner even commenting that the belief that nothing can be done is an important component of the frailty itself: "Frailty starts when a person thinks that nobody can help her" (Healthcare practitioner, Shaw et al. 2017, p. 19).

Others suggested that intervention was still worthwhile even if frailty itself could not be reversed: "I don't think you can prevent it but you can aid them to be the best they can be" (Social care practitioner, Shaw et al. 2017, p. 21). This lack of acceptance of the malleability of frailty was less common amongst the smaller group of participants in the policymakers study (Gwyther et al. 2018). Nevertheless the policymakers were aware that there was still a significant knowledge gap amongst healthcare practitioners in particular, with a concern raised that although geriatricians were aware of the issue, many other professionals who come into contact with frail older people every day were still unaware that frailty can be treatable:

> most of the professionals who work with adults and older people will be coming across frailty every day. They might not recognise what they can do about it but I think that they're aware of it as a challenge. I don't think they're fully aware of what the possibilities are. (Policy maker, Gwyther et al. 2018, p. 3)

This was seen as a significant challenge for them as people who are in charge of addressing frailty management: "Currently, the awareness around frailty is poor, then whatever we talk about afterwards is not going to happen, until we address this primary deficiency." (Gwyther et al. 2018)

However, even amongst this group of very well-informed participants, there was still some suggestion of uncertainty around the treatability of frailty, this time, from just one of the healthcare policymakers, with their willingness for screening for frailty being based on being able to work out who not to treat: "So it is good to know which patients they are so we do not undertake unnecessary medical interventions" (Gwyther et al. 2018, p. 7).

The Role of Health Psychology Theory

Putting this issue into the language of health psychology theory, addressing "outcome expectancies", or beliefs in the benefits of intervention and in the potential to actually have an impact on levels of frailty, is clearly a priority for interventions and for healthy lifestyle behaviour change. Another important component of positive health behaviour is one's belief in one's own capacity (or the capacity of one's organisation) to perform the required behaviours, that is, self-efficacy. Examples would be the belief in one's ability to do the prescribed exercise or make the dietary changes on the part of the older adults, or to provide appropriate support on the part of the health or social care practitioners.

Thus, it was clearly identified that a first task, as an important component of any frailty intervention, is to address belief in the possibility that frailty can be modified amongst both older adults and intervention providers. Using health psychology theory enables understanding and development of interventions that include such components. For example, a Theory of Planned Behaviour (TPB) approach (e.g., Ajzen 1991) would suggest that addressing attitudes towards the behaviour, that is, addressing the beliefs that a health behaviour will lead to certain outcomes, is one important way to have an impact on the intention to engage in a healthy behaviour. In our example of frailty, this would be addressing beliefs that a course of action could improve one's frailty. The TPB would also suggest that subjective norms, our motivation to comply with the views of others (e.g., one's GP, family, peers), would also have an impact and could be addressed. Perceived behavioural control is the final component in the TPB, which refers to one's belief or confidence in being able to perform the behaviour. This is related to self-efficacy but also to one's own perceived barriers. Together, the three components have an impact on intention to perform the health-related behaviour, for example, to do prescribed exercises.

Another useful model is the Health Action Process Approach (HAPA – Schwarzer 2008). As illustrated in Fig. 7.1, this includes expectancies of outcomes contingent on the health behaviour (e.g., beliefs in the effectiveness of an intervention for frailty), as well as self-efficacy, but also recognises that health behaviour and the development and maintenance of new habits and behaviours involve more than just the intention-behaviour

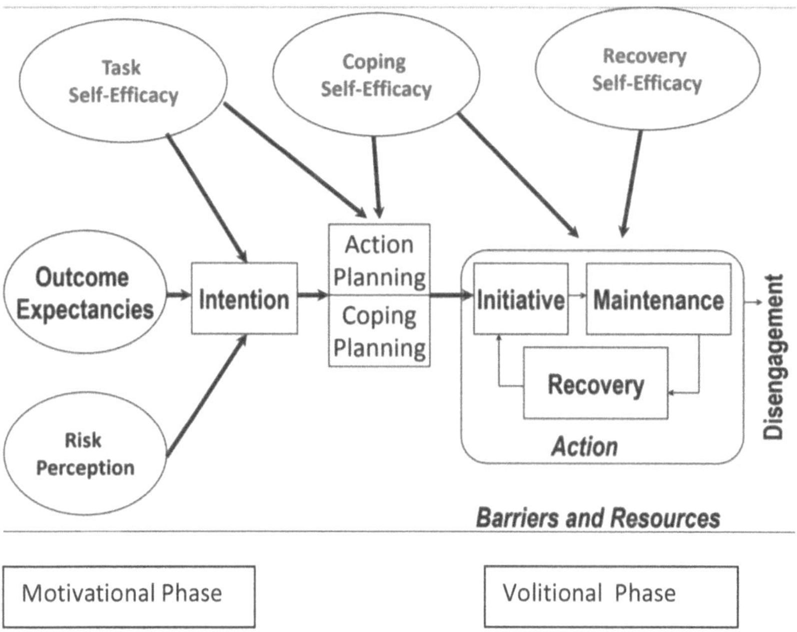

Fig. 7.1 The health action process approach (Schwarzer 2008)

link as suggested by the TPB. This model includes both the motivation for the behaviour and a volitional phase that includes maintenance self-efficacy, recovery self-efficacy, action planning, and coping planning. Action planning involves planning how, where, and when the behaviour will happen and is involved in the initiation of the behaviour, but coping planning is involved in both the initiation and the maintenance, based on the assumption that barriers will need to be overcome and coping with setbacks or reversions to previous behaviour is important. Likewise, maintenance self-efficacy refers to one's belief in one's own ability to perform the behaviour in an ongoing way, whereas recovery self-efficacy is belief in one's own ability to overcome setbacks or lapses.

Re-conceptualising frailty interventions as "behaviour change" as opposed to treatments in which the patient is a passive recipient of a prescription or procedure is an important role for the health psychologists involved in such programmes and perhaps requires a shift in mindset for medical practitioners used to referring patients for treatments or to writing prescriptions. Health psychologists are psychology specialists in the understanding of health behaviour and of enabling health behaviour change, with expertise in determining the sources of health behaviour issues such as lack of involvement in interventions or adherence to programmes or treatments. Specifically, they use theory-based strategies developed from research evidence and combine that with service provider and user perspectives to design behaviour change interventions appropriate to the user group, the target behaviour, and the healthcare or community environment, often working alongside medical practitioners. In this way, the behaviour required to make the most of the "prescription" from a medical practitioner or suggested by research evidence becomes more likely to improve health outcomes and have the effect the underlying basic science suggests it should have.

Research Example 2: Realist Methodology and the Context of Interventions

In combination with such background in health psychology, the outcomes of the array of studies from the FOCUS project were synthesised into a Realist Review. This type of review is aimed specifically at determining what works for whom and in what circumstances (Pawson et al. 2005). The aim was to produce evidence-based recommendations for intervention structures that would contribute to eventual international guidelines on frailty intervention (Gwyther et al. 2018; Marcucci et al. submitted). A realist process enables us to begin to understand the conditions or circumstances (the context, C) in which an intervention may work (or not work)—what are the conditions that may trigger the mechanism (M), or the process of how an intervention works, and leads to specified outcomes (O). Our study used the reviews of the literature on frailty from the FOCUS project to construct Context-Mechanism-Outcome (CMO) configurations. These theories were then tested out with our real-world data. Evidence from both the qualitative and quantitative components was used to support or refute CMOs in order to produce recommendations for frailty interventions. An example of a CMO theory is provided in Fig. 7.2.

This realist process concluded that initial frailty levels, presence or absence of specific deficits, and full person, organisational, and policy contexts should be taken into account in intervention design. Importantly, it was clear that strategies to enhance the social and psychological aspects

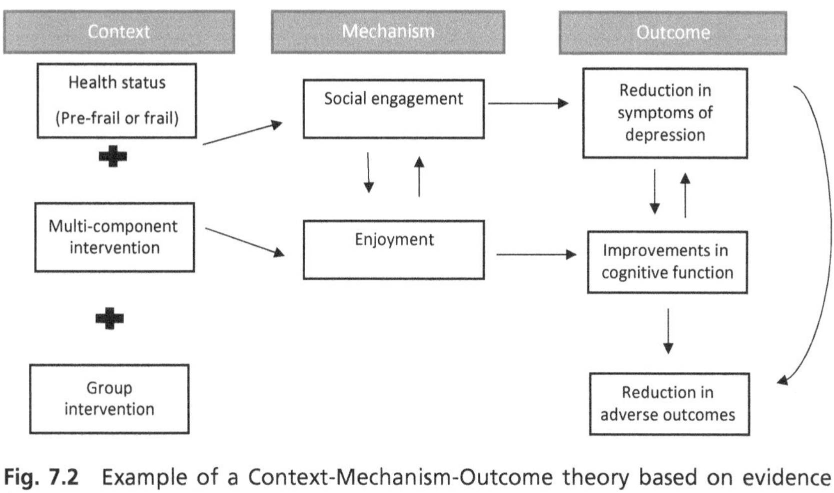

Fig. 7.2 Example of a Context-Mechanism-Outcome theory based on evidence from Gwyther et al. (2018)

of interventions and overcome environmental barriers should be included even in physically focused interventions, in order to improve adherence and intervention success. That is, it was clear in this work that strategies to improve an individual's social, environmental, and personal psychological resilience are just as important as the underlying science of the intervention strategy.

Resilience

The concept of resilience in the context of frailty is often described simply in physiological terms. Physiological resilience refers to the loss of reserve and the ability of our bodily systems to repair themselves, to respond to trauma and infection, and to maintain homoeostasis in bodily systems. Some authors (Varadhan et al. 2008) have suggested that this reduced capacity to maintain homoeostasis and the consequent increased vulnerability to internal and external stressors is due to the critical accumulation of deregulation in important signalling pathways, and others have suggested a reduced capacity to compensate for age-related cumulative damage at molecular and cellular levels (Clegg et al. 2013). However, as we have seen in the chapter on Stress and Immunity in Ageing in this volume (Chap. 5), there is an important interaction between the physiological, the psychological, and the environmental in the regulation of these systems. Examples from the FOCUS qualitative work illustrate this issue. For example, in the meta-synthesis (D'Avanzo et al. 2017) and in other sources (Ebrahimi et al. 2012; Ekwall et al. 2012), a clear sense of the possibilities of "health despite frailty" emerge. Two contrasting quotations, both from older women who would have been assessed as physically frail by a medical practitioner illustrate this issue:

> There is nothing really wrong with my health, but then I get all these ailments. I have fallen five times and broken my arms and my leg and got a hip replacement and broken pelvis and all my ribs, I've broken everything, in fact. So that's how I am, my bones are a chaos, but I have no health problems just now. (Ekwall et al. 2012, p. 15)

> I had a fall and broke my arm, and that seemed to upset the whole system. The pain from the arm wasn't too bad, but it seemed to upset me. … Things just seem different now. (Nicholson et al. 2013, p. 1429)

While the first example illustrates resilience despite obvious physical frailty, the second suggests that some frail older adults are in a precarious state of depleted resilience in which the next stressor, even if it is not a particularly serious one, may "tip them over the edge" (Roland et al. 2011, pp. 10–11). The question of what makes for this psychological resilience, and what are its implications for quality of life components, for example, sense of control, autonomy, self-realisation, and pleasure (Hyde et al. 2003), is an important one for us as psychologists, leading us to the concept of psychological frailty. Two components of psychological frailty that have been considered in the literature are cognitive frailty and mood and motivational frailty.

Cognitive Frailty The definition of cognitive frailty is very similar to that for Mild Cognitive Impairment (MCI) but in the context of physical frailty. That is, there are clear cognitive deficits that are at the level where a person may compensate quite well in everyday life, leading to little obvious impairment in activities of daily living (Kelaiditi et al. 2013). Using this commonly accepted definition, there should be no clear signs of organic damage (although there may be underlying undetected issues) and no dementia, with little cognitive reserve, such that a stressor such as an acute illness may result in confusional states. Cognitive frailty is present in 22–45% of the people with physical frailty, with Boyle, Buchman, Wilson, Leurgans, and Bennett (2010) demonstrating, in a longitudinal study, that physical frailty increases the risk of cognitive frailty over time. The direction of this relationship is suggestive of a common physical cause. While there is plentiful evidence of the relationship between different chronic illnesses and cognitive change in older age (e.g., cardiovascular disease, diabetes, chronic obstructive pulmonary disorders, for a review see Holland and Rabbitt 1991), a direct relationship between the common physical manifestations of frailty such as sarcopenia (muscle wastage) and cognitive symptoms is not well evidenced.

Mood and Motivational Frailty Mood and motivational frailty, (feelings of inability to cope or compensate for difficulties, low energy and motivation, e.g., for social, physical or self-care activities) is closely related to depression, which is common in frailty. The prevalence of significant depression is higher in a population of frail older adults than in the general age-matched population with an estimate of 4–16% of frail older adults aged 60 being depressed compared with 1–5% in the general older population (Collard et al. 2014). Prevalence for frail people aged over 75 is even higher, at 35% (Jürschik et al. 2012), although studies show higher percentages when simple co-morbidity is examined—46–57% (Collard et al. 2014). Collard et al. also demonstrated that the prevalence of frailty was three times higher in depressed than in non-depressed individuals after controlling for baseline characteristics such as age and sex. The risk of frailty increases longitudinally if depression is present (Chang et al. 2010), and the risk of depression is increased longitudinally in those who are physically frail or pre-frail at baseline (Feng et al. 2014) with Mezuk et al. (2012) suggesting a bi-directional relationship. Collard et al. (2014) carefully separated out some of the overlapping symptoms of depression and physical frailty (e.g., feelings of exhaustion and low activity) and still found a significant association between the two issues. They suggested that for some frail older adults, the confluence of depression and physical frailty may not be distinct syndromes but may indeed be a single construct, underlining the proposal that measures of frailty severity do need to include depression severity. This was illustrated by Brown et al. (2014), who demonstrated that the severity of individual frailty criteria such as slowing of gait speed, reduction in muscle strength, or self-reported exhaustion was worse in a depressed as opposed to a non-depressed frail group. This severity impact also increased the impact of frailty on outcomes such as mortality. Importantly, in a review of this relationship between frailty and depression, Buigues et al. (2015) cited replicated evidence that the increased prevalence of depressive symptoms in the context of frailty is maintained after adjusting for sociodemographic factors, co-morbidities, and functional disabilities. Thus it seems that the common co-existence of depression and frailty is not simply related to potentially co-existing factors that may be assumed to reduce reserve and resilience and therefore add to the likelihood of depression in the context of frailty.

Other authors have suggested that psychological frailty, particularly factors leading to depression, may be associated with the loss of agency and self-efficacy that may come with physical and functional impairments and loss of ability to maintain independently one's own normal activities. This is suggestive of possible mechanisms: those who become frail may be doing so because depression is causing them to withdraw from daily activities or have a poor nutritional intake and vice versa—those who are frail may also have difficulty engaging in physical or social activities increasing the risk of depression. Likewise anxiety may result from increasing functional limitations, but persistent anxiety and stress are also risk factors for accumulation of further co-morbidities (see also Chap. 5).

However, mood and motivational frailty is described as feelings of anxiety and inability to cope with problems in addition to depression, experienced in parallel with physical frailty symptoms (Fitten 2015), a description that again highlights the concept of an accumulation of deficits that may overwhelm one's own personal resources and ability to cope—that is, psychological resilience.

Psychological Resilience

Historically, research into adversity has focused on illness and psychopathology despite philosophical and psychological recognition of the human capacity for growth as a result of suffering (Calhoun and Tedeschi 2006). It is only recently in Western society that adversity has been examined from a positive psychological perspective, focusing on the human capacity to achieve optimal levels of health and well-being (Joseph and Linley 2005; Yu et al. 2014). It is from this paradigm shift towards a positive psychological standpoint that the development of growth from adversity and subsequent resilience has emerged (Joseph and Linley 2005). According to Joseph and Linley (2006), resilience is primarily recognisable through three components: (1) an enhancement in positive relationships with family and friends, as well as increased levels of compassion and empathy towards others; (2) increased levels of internal strength and wisdom, with a recognition of limitations and vulnerabilities; and (3) a greater appreciation for everyday life. Furthermore, upon self-reflection, the adversity is viewed as a valuable learning experience, with these authors proposing that resilience is a response to adversity. The adversity can vary from severe events such as cancer, bereavement, disability, and so on to lesser problems such as minor injuries or physical and cognitive decline and chronic illness (MacLeod et al. 2016). However, the process of overcoming adversity and developing resilience is neither linear nor definitive (Linley and Joseph 2005), and merely experiencing the adversity is not sufficient to trigger a response. The individual must perceive the adversity as an obstacle to overcome (Valdez and Lilly 2014). Furthermore, there is no guarantee the response provided to the adversity will be a positive resilience-promoting action. For example, focusing on negative intrinsic perceptions of life threat, uncontrollability, and helplessness is likely to result in maladaptive coping responses that not only inhibit the development of resilience but promote negative psychological adjustment and psychopathological development (Joseph and Linley 2005). However, if responses such as positive reinterpretation, adaptive coping, and effortful rumination are implemented, the adversity is likely to be overcome, resulting in resilience developing (Valdez and Lilly 2014). Therefore, by promoting factors associated with positive adaptive coping and minimising risk and impact of adversity, the capacity and self-belief to overcome adversity in individuals is greater and therefore increases the likelihood of developing resilience against stressors (Greenberg 2006). Indeed, such principles have been utilised and implemented in child and adolescent resilience-building interventions to promote psychological well-being such as the Promoting Alternative Thinking Strategies (PATHS) curriculum (Greenberg 2006). Whilst improving resilience in older adults remains theoretical (MacLeod et al. 2016), the emergence of effective resilience-building interventions confirms that resilience can be promoted.

As seen above, a prevalent conception is that resilience and frailty are polar opposites on the same dimension, that is, to be resilient is to not be frail, one cannot be resilient and frail simultaneously. However, as we have seen in our qualitative data, this is not only possible but common. Kuh (2007) defined resilience as "the capacity for adaptation in the face of ever-changing environmental challenges" (p. 719) and separated out the concepts of social and psychological resilience from that of biological or physiological resilience. This concept is useful in explaining the differences between the two frail older women illustrated above (Ekwall et al. 2012; Nicholson et al. 2013). To conclude that resilience is the absence of frailty overlooks the key attribute of resilience which is to regain lost mental well-being and physical function and not just maintain current capabilities (Herrman et al. 2011). Therefore, in the context of frailty, it is more accurate to define resilience as the presence of healthy and adaptive behaviour, rather than as the absence of illness, which may act as protective factors against further frailty development or against negative outcomes of frailty such as depression and poor quality of life.

Thus, a useful way to define resilience would be as the ability to cope and adapt despite the presence of frailty (Ebrahimi et al. 2012). Resilience despite frailty is a recently developed notion. Research examining this link is predominantly theoretical (Luthar and Cicchetti 2007). For example, research has indicated that resilience is a key aspect of helping older adults adjust to hardships and that it can be promoted by maintaining strong social ties and good mental and physical health (Wells 2012). Despite this, the application of theory in practice is scarce (Luthar and Cicchetti 2007; MacLeod et al. 2016), which means that evidence on interventions to promote resilience as opposed to reduce or prevent frailty is very limited. Furthermore, the long-term benefits of maintaining high resilience in relation to frailty are yet to be fully researched. For example, does high resilience reduce or prevent further frailty? Are prognostic outcomes of frailty improved if resilience is high? And can resilience be promoted in older adults who are already frail despite consistent conclusions in research highlighting the importance of early intervention?

Research Focus 3: Strategies to Prevent or Respond to Frailty

In our meta-synthesis on stakeholders' views and experiences (D'Avanzo et al. 2017), a second theme identified was people's *strategies to prevent or respond to frailty*. Adaptive coping was very apparent, with people describing their natural Selective Optimisation and Compensation

(SOC) strategies (Baltes 1997). Baltes and Baltes (1990) used the concept of resilience to suggest a definition of "successful" ageing which consisted of a positive balance between gains and losses such that they not only coped with declines, for example, using compensatory strategies and selective optimisation, but that they also continued to develop themselves in different ways.

The response to losses, deficits, or functional impairment, rather than the loss itself, is indicated as what makes the difference between people who manage to maintain well-being and those who struggle. Baltes (1997) suggested "loss based selection" as a way to explain how individuals maximise their potential gains and minimise potential losses by adaptive selection of goals and optimisation of their route towards these selected goals. It was apparent in our qualitative data that this involved a degree of acceptance—people were selecting what they judged to be changeable and unchangeable and focusing on what they could control and also on their goals. For one woman with visual and mobility impairments cited by Hjaltadottir and Gustafsdottir (2007), goals clearly included maintenance of lifelong personal development, in the context of acceptance of limitations: "If you learn to accept things as they are, use the time that otherwise would be empty and listen to tapes, then you can go on learning until you die" (p. 52).

There were examples of personal strategies to "keep frailty at a distance" (Horder et al. 2013) but also examples of how the frail older person maintained personal control and self-efficacy in the context of having to accept help from formal and informal care providers, such as for one man who could no longer tend his own garden:

> Well, that's difficult for me (working in the garden) so I'm now training a whole lot of junior gardeners. You'll be surprised, I'm training Jackie and anybody that comes now who's not a gardener; I'm training a whole new generation, because you see I have to remember every day. (Nicholson et al. 2013, p. 1177)

To many older adults, keeping frailty at a distance involved keeping a harmony or balance in everyday life, with keeping a routine and keeping social connections seen as important components of resilience, as this woman illustrates in her appreciation of her ability to be connected:

> It's a matter of luck that I can still do so much, that my mind is still clear, that I can say something, that I can talk to them (caregivers). … I can chat with my children about the weather, about what's going on in the world. … I can take part in things, to a certain extent! (Claassens et al. 2014, p. 165)

These examples underline the concepts of SOC, but also of resilience that includes acceptance as well as maintenance of control. Other examples show clear strategic planning both to compensate and to reduce further stressors, such as by sticking to routines and habits and to use whatever resources are available to maintain well-being. As De Alfieri and Borgogni (2010) commented, "resilience can be described by viewing constant competence under stress, recovery from a dramatic event, favorable outcomes regardless of high risk status, and ability to build and increase the capacity for learning and adaptation" (p. 602). Crucial to this discussion is the issue of whether resilience is a stable personality trait or whether it is adaptable, learnt, context dependent, and/or modifiable. Our evidence suggests that coping strategies are an important component of resilience and that finding ways to maintain autonomy may be a potential mechanism for improving the outcomes of frailty. White et al. (2010) reported that depression, satisfaction with life, and functional independence were correlated with resilience, while other writers have described resilience as consisting of internal personal assets and external or social assets (e.g., Staudinger et al. 1993), leading to the suggestion of various avenues to use to examine just how much resilience can be supported or developed. Importantly, to enable this, we need good measures of resilience, with significant progress made in developing measures specifically for use with older populations (e.g., van Abbema et al. 2015).

Conclusion

The concepts of resilience and frailty cannot be seen as entirely opposite ends of a state of balance, or health, given clear evidence of resilience despite frailty. The construct of resilience seems to consist of both internal and external assets, leading us to link it to the recommendations from the realist synthesis of outcomes of the FOCUS studies. There we found indications that social, psychological, environmental, and organisational support is necessary for the success of interventions for frailty. The necessity of including theory-based intervention design concepts, such as addressing maintenance, self-efficacy, coping, or outcome expectancies can also be seen as supporting and developing resilience. In conclusion, interventions that address resilience as well as directly aim to address physiological and/or cognitive frailty are likely to have more success.

References

Ajzen, I. (1991). The Theory of Planned Behavior. *Organizational Behavior and Human Decision Processes, 50*(2), 179–211.

Apóstolo, J., Cooke, R., Bobrowicz-Campos, E., Santana, S., Marcucci, M., Cano, A., et al. (2017). Predicting Risk and Outcomes for Frail Older Adults: An Umbrella Review of Frailty Screening Tools. *JBI Database of Systematic Reviews and Implementation Reports, 15*(4), 1154–1208.

Auyeung, T. W., Lee, J., Kwok, T., & Woo, J. (2011). Physical Frailty Predicts Future Cognitive Decline—A Four-Year Prospective Study in 2737 Cognitively Normal Older Adults. *Journal of Nutrition, Health and Aging, 15*(8), 690–694.

Baltes, P. B. (1997). On the Incomplete Architecture of Human Ontogeny: Selection, Optimization, and Compensation as Foundation of Developmental Theory. *American Psychologist, 52*, 366–380.

Baltes, P. B., & Baltes, M. M. (1990). Psychological Perspectives on Successful Aging: The Model of Selective Optimization with Compensation. In P. B. Baltes & M. M. Baltes (Eds.), *Successful Aging: Perspectives from the Behavioral Sciences* (pp. 1–34). New York: Cambridge University Press

Bleijenberg, N., Ten Dam, V. H., Steunenberg, B., Drubbel, I., Numans, M. E., De Wit, N. J., & Schuurmans, M. J. (2013). Exploring the Expectations, Needs and Experiences of General Practitioners and Nurses Towards a Proactive and Structured Care Programme for Frail Older Patients: A Mixed-Methods Study. *Journal of Advanced Nursing, 69*(10), 2262–2273.

Bowling, A. (2009). Predictors of Mortality Among a National Sample of Elderly Widowed People: Analysis of 28-year Mortality Rates. *Age and Ageing, 38*(5), 527–530.

Boyle, P. A., Buchman, A. S., Wilson, R. S., Leurgans, S. E., & Bennett, D. A. (2010). Physical Frailty Is Associated with Incident Mild Cognitive

Impairment in Community Based Older Persons. *Journal of the American Geriatrics Society, 58*(2), 248–255.

Brown, P. J., Roose, S. P., Fieo, R., Liu, X., Rantanen, T., Sneed, J. R., et al. (2014). Frailty and Depression in Older Adults: A High-Risk Clinical Population. *The American Journal of Geriatric Psychiatry, 22*(11), 1083–1095.

Buigues, C., Padilla-Sánchez, C., Garrido, J. F., Navarro-Martínez, R., Ruiz-Ros, V., & Cauli, O. (2015). The Relationship Between Depression and Frailty Syndrome: A Systematic Review. *Aging & Mental Health, 19*, 762–772.

Cadore, E., Casas-Herrero, A., Zambom-Ferraresi, F., Idoate, F., Millor, N., Gómez, M., et al. (2014). Multicomponent Exercises Including Muscle Power Training Enhance Muscle Mass, Power Output, and Functional Outcomes in Institutionalized Frail Nonagenarians. *Age (Dordr), 36*(2), 773–785.

Calhoun, L. G., & Tedeschi, R. G. (2006). The Foundations of Posttraumatic Growth: An Expanded Framework. In L. G. Calhoun & R. G. Tedeschi (Eds.), *Handbook of Posttraumatic Growth: Research and Practice* (pp. 3–23). Mahwah: Erlbaum

Cameron, I. D., Fairhall, N., Langron, C., Lockwood, K., Monaghan, N., Aggar, C., et al. (2013). A Multifactorial Interdisciplinary Intervention Reduces Frailty in Older People: Randomized Trial. *BMC Medicine, 11*(1), 1–10.

Chan, D., Tsou, H. H., Yang, R. S., Tsauo, J. Y., Chen, C. Y., Hsiung, C. A., & Kuo, K. N. (2012). A Pilot Randomized Controlled Trial to Improve Geriatric Frailty. *BMC Geriatrics, 12*, 58.

Chang, S. S., Weiss, C. O., Xue, Q. L., & Fried, L. P. (2010). Patterns of Comorbid Inflammatory Diseases in Frail Older Women: The Women's Health and Aging Studies I and II. *The Journals of Gerontology Series A: Biological Sciences and Medical Sciences, 65A*(4), 407–413.

Claassens, L., Widdershoven, G. A., Van Rhijn, S. C., Van Nes, F., Broese van Groenou, M. I., Deeg, D. J., & Huisman, M. (2014). Perceived Control in Health Care: A Conceptual Model Based on Experiences of Frail Older Adults. *Journal of Aging Studies, 31*, 159–170.

Clegg, A., Young, J., Iliffe, S., Rikkert, M. O., & Rockwood, K. (2013). Frailty in Elderly People. *Lancet, 381*, 752–762.

Collard, R. M., Boter, H., Schoevers, R. A., & Voshaar, R. C. O. (2012). Prevalence of Frailty in Community-Dwelling Older Persons: A Systematic Review. *Journal of the American Geriatrics Society, 60*(8), 1487–1492.

Collard, R. M., Comijs, H. C., Naarding, P., & Voshaar, R. C. A. (2014). Physical Frailty: Vulnerability of Patients Suffering from Late-Life Depression. *Aging & Mental Health, 18*(5), 570–578.

Craig, P., Dieppe, P., Macintyre, S., Michie, S., Nazareth, I., & Petticrew, M. (2008). Developing and Evaluating Complex Interventions: The New Medical Research Council Guidance. *British Medical Journal, 337*, a1655.

D'Avanzo, B., Shaw, R., Riva, S., Apostolo, J., Bobrowicz-Campos, E., Kurpas, D., et al. (2017). Stakeholders' Views and Experiences of Care and Interventions for Addressing Frailty and Pre-frailty: A Meta-synthesis of Qualitative Evidence. *PLoS One, 12*(7), e0180127.

De Alfieri, W., & Borgogni, T. (2010). Through the Looking Glass and What Frailty Found There: Looking For Resilience in Older Adults. *Journal of the American Geriatrics Society, 58*(3), 602–603.

Ebrahimi, Z., Wilhelmson, K., Moore, C. D., & Jakobsson, A. (2012). Frail Elders' Experiences with and Perceptions of Health. *Qualitative Health Research, 22*(11), 1513–1523.

Ekwall, A., Hallberg, I. R., & Kristensson, J. (2012). Compensating, Controlling, Resigning and Accepting-Older Person's Perception of Physical Decline. *Current Aging Science, 5*(1), 13–18.

Faes, M. C., Reelick, M. F., Joosten-Weyn Banningh, L. W., Gier, M., Esselink, R. A., & Olde Rikkert, M. G. (2010). Qualitative Study on the Impact of Falling in Frail Older Persons and Family Caregivers: Foundations for an Intervention to Prevent Falls. *Aging & Mental Health, 14*(7), 834–842.

Feng, L., Nyunt, M. S., Feng, L., Yap, K. B., & Ng, T. P. (2014). Frailty Predicts New and Persistent Depressive Symptoms Among Community-Dwelling Older Adults: Findings From Singapore Longitudinal Aging Study. *Journal of the American Medical Directors Association, 15*(1), 76.e7_76.e12.

Ferrucci, L., Guralnik, J. M., Studenski, S., Fried, L. P., Cutler, G. B., Walston, J. D., & The Interventions on Frailty Working Group. (2004). Designing Randomized, Controlled Trials Aimed at Preventing or Delaying Functional Decline and Disability in Frail, Older Persons: A Consensus Report. *Journal of the American Geriatrics Society, 52*(4), 625–634.

Fitten, L. J. (2015). Psychological Frailty in the Aging Patient. In R. A. Fielding, C. Sieber & B. Vellas (Eds.), *Frailty: Pathophysiology, Phenotype and Patient Care, Nestlé Nutrition Institute Workshop Series* (Vol. 83, pp. 45–53). Basel: Karger.

Fried, L. P., Tangen, C. M., Walston, J., Newman, A. B., Hirsch, C., Gottdiener, J., et al. (2001). Frailty in Older Adults: Evidence for a Phenotype. *Journals of Gerontology A Biological Sciences & Medical Sciences, 56*(3), M146–M157.

Fried, L. P., Ferrucci, L., Darer, J., Williamson, J. D., & Anderson, G. (2004). Untangling the Concepts of Disability, Frailty, and Comorbidity: Implications For Improved Targeting and Care. *Journals of Gerontology A Biological Sciences & Medical Sciences, 59*(3), 255–263.

Gill, T. M., Baker, D. I., Gottschalk, M., Peduzzi, P. N., Allore, H., & Byers, A. (2002). A Program to Prevent Functional Decline in Physically Frail, Elderly Persons Who Live at Home. *New England Journal of Medicine, 347*(14), 1068–1074.

Greenberg, M. T. (2006). Promoting Resilience in Children and Youth. *Annals of the New York Academy of Sciences, 1094*, 139–150.

Gwyther, H., Cooke, R., Shaw, R., Marcucci, M., Cano, A., & Holland, C. (2017). Perceptions and Experiences of Frailty Interventions: Quantitative and Qualitative Results from a Survey of Partners Within the European Innovation Partnership on Active and Healthy Ageing (EIP-AHA). *Ageing and Society*, 1–25. https://doi.org/10.1017/S0144686X17000265

Gwyther, H., Shaw, R., Jaime-Dauden, E.-A., D'Avanzo, B., Kurpas, D., Bujnowska-Fedak, M., et al. (2018). Understanding Frailty: A Qualitative Study of European Healthcare Policy-makers' Approaches to Frailty Screening and Management. *BMJ Open, 8*(1), e018653.

Gwyther, H., Bobrowicz-Campos, E., Apostolo, J., Marcucci, M., Cano, A., & Holland, C. (submitted). *A Realist Review to Understand the Efficacy and Outcomes of European Frailty Interventions in the Country, Healthcare Organisational and Patient Contexts.*

Herrman, H., Stewart, D. E., Diaz-Granados, N., Berger, E. L., Jackson, B., & Yuen, T. (2011). What Is Resilience? *Canadian Journal of Psychiatry, 56*(5), 258–265. https://doi.org/10.1177/070674371105600504

Hjaltadottir, I., & Gústafsdottir, M. (2007). Quality of Life in Nursing Homes: Perception of Physically Frail Elderly Residents. *Scandinavian Journal of Caring Sciences, 21*(1), 48–55.

Holland, C., & Rabbitt, P. (1991). The Course and Causes of Cognitive Change with Advancing Age. *Reviews in Clinical Gerontology, 1*(1), 81–96.

Holland, C. A., Everitt, P., Johnson, A., & Devi, R. (2008). The 'Healthy Passport' Intervention with Older People in an English Urban Environment: Effects of Incentives and Peer-Group Organisers in Promoting Healthy Living. *Ageing and Society, 28*(4), 525–549.

Holland, C., Cooper, Y., Shaw, R., Pattison, H., & Cooke, R. (2013). Effectiveness and Uptake of Screening Programmes for Coronary Heart Disease and Diabetes: A Realist Review of Design Components Used in Interventions. *BMJ Open, 3*(11), e003428.

Holland, C. A., Carter, M., Cooke, R., Leask, G., Powell, R., Shaw, R., et al. (2015). *Collaborative Research Between Aston Research Centre for Healthy Ageing (ARCHA) and the ExtraCare Charitable Trust* (Final Report). Available on http://www.aston.ac.uk/lhs/research/centres-facilities/archa/extracare-project/

Horder, H. M., Frandin, K., & Larsson, M. E. H. (2013). Self-respect Through Ability to Keep Fear of Frailty at a Distance: Successful Ageing from the Perspective of Community-Dwelling Older People. *International Journal of Qualitative Studies on Health and Well-being, 8*, 20194.

Hyde, M., Wiggins, R. D., Higgs, P., & Blane, D. B. (2003). A Measure of Quality of Life in Early Old Age: The Theory, Development and Properties of a Needs Satisfaction Model (CASP-19). *Aging & Mental Health, 7*, 186–194.

Joseph, S., & Linley, P. A. (2005). Positive Adjustment to Threatening Events: An Organismic Valuing Theory of Growth Through Adversity. *Review of General Psychology, 9*(3), 262–280.

Joseph, S., & Linley, P. A. (2006). Growth Following Adversity: Theoretical Perspectives and Implications for Clinical Practice. *Clinical Psychology Review, 26*, 1041–1053.

Jürschik, P., Nunin, C., Botigué, T., Escobar, M. A., Lavedán, A., & Viladrosa, M. (2012). Prevalence of Frailty and Factors Associated with Frailty in the Elderly Population of Lleida, Spain: The FRALLE Survey. *Archives of Gerontology and Geriatrics, 55*(3), 625–631.

Kelaiditi, E., Cesari, M., Canevelli, M., van Kan, A. G. O. P.-J., Gillette-Guyonnet, S., et al. (2013). Cognitive frailty: Rational and Definition from an (I.A.N.A./I.A.G.G.) International Consensus Group. *The Journal of Nutrition, Health & Aging, 17*, 726–734.

Keles, R. (2012). The Quality of Life and the Environment. *Procedia – Social and Behavioral Sciences, 35*, 23–32.

Kuh, D. (2007). A Life Course Approach to Healthy Aging, Frailty, and Capability. *The Journals of Gerontology Series A: Biological Sciences and Medical Sciences, 7*(1), 717–721.

Lally, F., & Crome, P. (2007). Understanding Frailty. *Postgraduate Medical Journal, 83*(975), 16–20.

Lang, P. O., Michel, J. P., & Zekry, D. (2009). Frailty Syndrome: A Transitional State in a Dynamic Process. *Gerontology, 55*, 539–549.

Levers, M. J., Estabrooks, C. A., & Kerr, J. C. R. (2006). Factors Contributing to Frailty: Literature Review. *Journal of Advanced Nursing, 56*(3), 282–291.

Lindhardt, T., Hallberg, I. R., & Poulsen, I. (2008). Nurses' Experience of Collaboration with Relatives of Frail Elderly Patients in Acute Hospital Wards: A Qualitative Study. *International Journal of Nursing Studies, 45*(5), 668–681.

Linley, P. A., & Joseph, S. (2005). The Human Capacity for Growth Through Adversity. *The American Psychologist, 60*(3), 262–264.

Littlechild, R., & Glasby, J. (Eds.). (2000). *Older People as Participating Patients. User Involvement and Participation in Social Care*. London: Jessica Kingsley.

Liu, C. J., & Latham, N. K. (2009). Progressive Resistance Strength Training for Improving Physical Function in Older Adults. *Cochrane Database of Systematic Reviews, 3*, CD002759.

Luthar, S. S., & Cicchetti, D. (2007). The Construct of Resilience: Implications for Interventions and Social Policies. *Development and Psychopathology, 12*(4), 857–885.

MacLeod, S., Musich, S., Hawkins, K., Alsgaard, K., & Wicker, E. R. (2016). The Impact of Resilience Among Older Adults. *Geriatric Nursing, 37*(4), 266–272.

Marcucci, M., Damanti, S., Germini, F., Apostolo, J., Bobrowicz-Campos, E., Gwyther, H., Holland, C., Kurpas, D., Bujnowska, M., Santana, S., Nobili, A., D'Avanzo, B., Szwamel, K., Villacampa Fernández, P., Shaw, R., Cooke, R., & Cano, A. (Submitted) Clinical Guidelines to Inform Decision-Making and Implementation of Best Practices to Prevent, Delay or Reverse Frailty.

Mezuk, B., Lohman, M., Dumenci, L., & Lapane, K. L. (2012). Are Depression and Frailty Overlapping Syndromes in Mid and Late-life? A Latent Variable Analysis. *The American Journal of Geriatric Psychiatry, 21*, 560–569.

Morley, J. E. (2013). Frailty: A Time for Action. *European Geriatric Medicine., 4*(4), 215–216.

Ng, T. P., Feng, L., Nyunt, M. S. Z., Feng, L., Niti, M., Tan, B. Y., et al. (2015). Nutritional, Physical, Cognitive, and Combination Interventions and Frailty Reversal Among Older Adults: A Randomized Controlled Trial. *The American Journal of Medicine, 128*(11), 1225–1236 e1.

Nicholson, C., Meyer, J., Flatley, M., & Holman, C. (2013). The Experience of Living at Home with Frailty in Old Age: A Psychosocial Qualitative Study. *International Journal of Nursing Studies, 50*(9), 1172–1179.

Oakley, A., Strange, V., Bonell, C., Allen, E., Stephenson, J., & RIPPLE Study Team. (2006). Process Evaluation in Randomised Controlled Trials of Complex Interventions. *British Medical Journal, 332*, 413–416.

Pawson, R., Greenhalgh, T., Harvey, G., & Walshe, K. (2005). Realist Review – A New Method of Systematic Review Designed for Complex Policy interventions. *Journal of Health Services Research & Policy, 10*(suppl 1), 21–34.

Peterson, M. J., Giuliani, C., Morey, M. C., Pieper, C. F., Evenson, K. R., Mercer, V., et al. (2009). Physical Activity as a Preventative Factor for Frailty: The Health, aging, and Body Composition Study. *Journal of Gerontology, 64*(1), 61–68.

Puts, M. T. E., Shekary, N., Widdershoven, G., Heldens, J., & Deeg, D. J. H. (2009). The Meaning of Frailty According to Dutch Older Frail and Non-frail Persons. *Journal of Aging Studies, 23*(4), 258–266.

Rockwood, K., & Mitnitski, A. (2007). Frailty in Relation to the Accumulation of Deficits. *The Journals of Gerontology Series A: Biological Sciences and Medical Sciences, 62*(7), 722–727.

Rockwood, K., & Mitnitski, A. (2011). Frailty Defined by Deficit Accumulation and Geriatric Medicine Defined by Frailty. *Clinics in Geriatric Medicine, 27*(1), 17–26.

Rockwood, K., Mitnitski, A., Song, X., Steen, B., & Skoog, I. (2006). Long-term Risks of Death and Institutionalization of Elderly People in Relation to Deficit Accumulation at Age 70. *Journal of the American Geriatrics Society, 54*(6), 975–979.

Rodríguez-Mañas, L., Féart, C., Mann, G., Viña, J., Chatterji, S., Chodzko-Zajko, W., et al. (2013). Searching for an Operational Definition of Frailty: A Delphi Method Based Consensus Statement. The Frailty Operative Definition-Consensus Conference Project. *The Journals of Gerontology. Series A, Biological Sciences and Medical Sciences, 68*(1), 62–67.

Roland, K. P., Theou, O., Jakobi, J. M., Swan, L., & Jones, G. R. (2011). Exploring Frailty: Community Physical and Occupational Therapists' Perspectives. *Physical & Occupational Therapy in Geriatrics, 29*(4), 270–286.

Schwarzer, R. (2008). Modeling Health Behaviour Change: How to Predict and Modify the Adoption and Maintenance of Health Behaviours. *Applied Psychology. An International Review, 57*, 1–29.

Shaw, R. L., West, K., Hagger, B., & Holland, C. A. (2016a). Living Well to the End: A Phenomenological Analysis of Life in Extra Care Housing. *International Journal of Qualitative Studies on Health and Well-Being, 11*. https://doi.org/10.3402/qhw.v11.31100

Shaw, R. L., Holland, C., Pattison, H. M., & Cooke, R. (2016b). Patients' Perceptions and Experiences of Cardiovascular Disease and Diabetes Prevention Programmes: A Systematic Review and Framework Synthesis Using the Theoretical Domains Framework. *Social Science & Medicine, 156*, 192–203.

Shaw, R. L., Gwyther, H., Holland, C., Bujnowska-Fedak, M., Kurpas, D., Cano, A., et al. (2017). Understanding Frailty: Meanings and Beliefs About Screening and Prevention Across Key Stakeholder Groups in Europe. *Ageing and Society*, 1–30. https://doi.org/10.1017/S0144686X17000745

Staudinger, U. M., Marsiske, M., & Baltes, P. B. (1993). Resilience and Levels of Reserve Capacity in Later Adulthood: Perspectives from Life-Span Theory. *Development and Psychopathology, 5*, 541–566.

Theou, O., Stathokostas, L., Roland, K., Jakobi, J., Patterson, C., Vandervoort, A., & Jones, G. (2011). The Effectiveness of Exercise Interventions for the Management of Frailty: A Systematic Review. *Journal of Aging Research, 2011*, 569194.

Topinková, E. (2008). Aging, Disability and Frailty. *Annals of Nutrition & Metabolism, 52*(Suppl.1), 6–11.

UK CRC. (2003). *UK CRC Screening Pilot Evaluation Team/Evaluation of the UK Colorectal Cancer Screening Pilot: Final Report.* Public Health England, London.

Valdez, C. E., & Lilly, M. M. (2014). Posttraumatic Growth in Survivors of Intimate Partner Violence: An Assumptive World Process. *Journal of Interpersonal Violence, 30*(2), 215–231. https://doi.org/10.1177/0886260514533154

van Abbema, R., Bielderman, A., De Greef, M., Hobbelen, H., Krijnen, W., & van der Schans, C. (2015). Building from a Conceptual Model of the Resilience Process During Ageing, Towards the Groningen Aging Resilience Inventory. *Journal of Advanced Nursing, 71*(9), 2208–2219.

Varadhan, R., Seplaki, C. L., Xue, Q.-L., Bandeen-Roche, K., & Fried, L. P. (2008). Stimulus-response Paradigm for Characterizing the Loss of Resilience in Homeostatic Regulation Associated with Frailty. *Mechanisms of Ageing and Development, 129*, 666–670.

Waller, J., Jackowska, M., Marlow, L., & Wardle, J. (2012). Exploring Age Differences in Reasons for Nonattendance for Cervical Screening: A Qualitative Study. *BJOG: An International Journal of Obstetrics & Gynaecology, 119*(1), 26–32.

Wells, M. (2012). Resilience in Older Adults Living in Rural, Suburban, and Urban Areas. *Online Journal of Rural Nursing and Health Care, 10*(2), 45–54.

White, B., Driver, S., & Warren, A. M. (2010). Resilience and Indicators of Adjustment During Rehabilitation from a Spinal Cord Injury. *Rehabilitation Psychology, 55*(1), 23–32.

Young, H. M. (2003). Challenges and Solutions for Care of Frail Older Adults. *Online Journal Issues in Nursing, 8*(2), 46–69.

Young, A. C. M., Glaser, K., Spector, T. D., & Steves, C. J. (2016). The Identification of Hereditary and Environmental Determinants of Frailty in a Cohort of UK Twins. *Twin Research and Human Genetics, 19*(6), 600–609.

Yu, X., Lau, J. T. F., Zhang, J., Mak, W. W. S., Choi, K. C., Lui, W. W. S., et al. (2014). Social Support, Coping Strategies and Health-related Quality of Life Among Primary Caregivers of Stroke Survivors in China. *Journal of Clinical Nursing, 22*(15–16), 2160–2171.

Zuliani, G., Soavi, C., Maggio, M., De Vita, F., Cherubini, A., & Volpato, S. (2015). Counteracting Inflammation and Insulin Resistance with Diet and Exercise: A Strategy for Frailty Prevention? *European Geriatric Medicine, 6*(3), 220–231.

Part III

Community and Political Psychology

8

Co-creating Home and Community: Building Partnerships to Support Older Adults to Age-Well-in-Place

Judith Sixsmith, Mei Lan Fang, and Sarah Canham

Introduction

The provision of home and community supports can enable people to successfully age-in-place by supporting physical and mental health, social participation and independence, autonomy, and choice. One challenge to achieving positive ageing in place concerns the integration of place-based supports in the home and community environment as older adults transition into new homes. Sustainable solutions within a strong theoretical basis are needed; and in this case example, ageing in place and community psychology principles

J. Sixsmith (✉)
School of Nursing and Health Sciences, Dundee University, Dundee, Scotland
e-mail: j.sixsmith@dundee.ac.uk

M. L. Fang
The Urban Institute, Heriot-Watt University, Edinburgh, Scotland

S. Canham
Gerontology Research Centre, Simon Fraser University, Burnaby, BC, Canada
e-mail: scanham@sfu.ca

E. Peel et al. (eds.), *Psychologies of Ageing*,
https://doi.org/10.1007/978-3-319-97034-9_8

and values guide the implementation of this project with the full involvement of local communities, service organisations, and older adults themselves. Here, the diversity of older adults, their needs, wishes, and the ways in which they construct meaning in places are prioritised. Since partnership building is an important component of this process, we demonstrate in this chapter how shared interests and common goals of multiple stakeholder groups were enabled which built solidarity between partners as well as generated synergistic outcomes.

This chapter details the intricacies of developing partnerships with low-income older adults, local service providers (including property developers and property managers), and a housing society in the context of a Canadian housing redevelopment for older adults. We describe the partnership-building process which progressed through a series of steering committee meetings, interviews, photo-tours, deliberative dialogue workshops, community mapping workshops, a community 'walk along' session, and a feedback forum. The overarching aims of the research were to: (1) understand the challenges and opportunities experienced by older adults, service providers, and housing providers during the redevelopment; (2) identify how service providers and housing providers could best deliver age-friendly services and resources; and (3) determine actions that could be made to meet the needs of service providers and housing providers in order to help them better serve older adults. The research was conducted with the goal of co-creating equitable, meaningful housing solutions with older adults transitioning into affordable housing.

Ageing, Place, and Transitions

Past research has indicated the importance of place for older adults, particularly the home and community environment. Several studies have described how various aspects of wellbeing, including individual subjective experiences of wellbeing, social or relational and community wellbeing are influenced by the connection between people and the places which make up the matrix of their lives (Devine-Wright and Lyons 1997; Dixon and Durrheim 2000; Proshansky et al. 1983; Relph 1976; Sixsmith 1986; Tuan 1977; Twigger-Ross and Uzzell 1996). This early body of work has highlighted the complexity of the person-place relationship and paved the way for research on two mutually constituted theoretical understandings of older adults in place: ageing in place and sense of place.

The notion of ageing in place, a strong driver for articulating housing policy for older adults, is built on the knowledge that older adults would prefer to live in their own homes in communities of familiarity and is underpinned by the assumption that ageing in place is a positive experience. However, while ageing in place can operate as a positive force in the lives of some older adults, for others it can be inappropriate (Sixsmith and Sixsmith 2008). This may be especially so for older adults whose homes are difficult or expensive to upkeep and need modifications as their functional abilities decline (Hwang et al. 2011). Ageing in place is also difficult to sustain when the local community no longer serves the psychosocial, environmental, or service needs and desires of older adults (Woolrych and Sixsmith 2018). For example, gentrification and urban renewal efforts can create situations whereby older adults are priced out of their communities, along with local health and social care services as popular of out-of-town corporations replace local shops. In such cases, contemporary urban cities can be viewed as 'unfriendly' or even 'hostile' to older adults, acting as a barrier to accessing social, economic, and civic opportunities (Scharf et al. 2005). Such unfriendliness can discourage active ageing, putting older adults at risk of isolation and loneliness (Buffel and Phillipson 2012). In response to this, the 'age-friendly cities' and subsequent 'age-friendly communities' agendas have developed to encourage active ageing (Davies and Kelly 2014). The age-friendly cities agenda has called for inclusive design enabling people of all ages to participate in communities and cities which thoughtfully cater to their needs. This involves enabling connections to people who are important to them, keeping as active and healthy as possible, and enjoying positive wellbeing even into very old ages. In this respect, the World Health Organisation (WHO 2007) age-friendly guide presents a framework of eight domains of age-friendliness: the built environment, transport, housing, social participation, respect and social inclusion, civic participation and employment, communication, and community support and health services. In working towards age-friendly environments, best practice guidelines have been developed to support a range of age-friendly features including walkability, housing conditions, and the design of inclusive outdoor spaces (Housing our Ageing Population: Plan for Implementation- HAPPI2 2012; Inclusive Design For Getting Outdoors-IDGO 2012). Through such design, it may be possible, as Golant (2015) has suggested, to strive for 'ageing in *the right* place'. This emphasises the fit between people and place, recognising that, as Oswald and Wahl (2003) argue, the accumulated experiences of home and community are particularly important as a person ages.

Nevertheless, simply changing the physical fabric of homes and communities is not sufficient to create age-inclusive environments, especially when places hold strong social, psychological, and emotional meanings and bonds for people (Manzo and Perkins 2006). Memories, social experiences, learnt routines, and habits all contribute to the ways persons relate to their local community and develop a sense of place belonging. A strong sense of place results from having access to supports for active participation and opportunities to build and sustain social networks and assume meaningful roles in the community (Seamon 2014). In contrast, a feeling of displacement or 'placelessness' (Relph 1976) is associated with alienation, isolation, and loneliness, often resulting in adverse health and wellbeing outcomes, particularly amongst vulnerable older adults (Lewicka 2013). This suggests that the creation of age-friendly urban environments that support sense of place is integral to successful ageing, ensuring that older adults can continue to make a positive contribution in old age, delaying the need for institutional care and reducing health and social care cost.

In attempting to prevent the problems associated with dislocation of place in later life (Sixsmith et al. 2018), it is advisable for those concerned with the social participation, health, and wellbeing of older adults to work in partnership with older adults so that their experiences of living in place and past relocations are taken into account. Partnering with older adults can help planners, architects, and other involved professionals to build places and communities that are inclusive, age-friendly, and supportive of health and wellbeing (Sixsmith et al. 2017). Working with older adults in collaborative partnership is more intensive than simply collecting data from older adults to inform decisions. Instead, working in partnership places older adults in decision-making roles and enables their voices to be heard. However, involvement can happen in very many different ways, following different principles, processes, and practices. In one recent example described later in this chapter, partnership working with older adults occurred during the redevelopment of a housing project for older, low-income adults, which occurred in parallel to a research project. A community psychology basis guided the research principles, processes, and practices so that issues of social justice and equity were forefronted. Thus, the research prioritised a social justice agenda, challenging status quo notions of how age-friendly places are constructed and emphasising the need to make a positive difference to the lives of more marginalised, low-income older adults. Below we discuss the key principles of community psychology, actualised in Community-Based Participatory approach, that guided this work and trace how they played a key part in shaping the research, the partnership process, and development practices through a case study of low-income older adults caught in a forced relocation situation.

The Approach: Community Psychology and Community-Based Participatory Research

The project's Community-Based Participatory Research (CBPR) approach, described more fully by Fang et al., was underpinned by the principles, assumptions, and values of community psychology. Community psychology offered an appropriate framing for this work since the issue of housing for older adults inherently involves (1) a grounding in community and because the work would contribute to the (2) production of social and individual change, (3) requiring the inclusion of the voice of low-income older people who are often marginalised within social systems. Community psychology can be understood as founded on a series of key interlinked principles (see Campbell and Murray 2004; Kagan et al. 2011): social justice, equity, *conscientiousization*, participation and collaboration, attention to power dynamics, and a plurality of thought and action based on viewing social (or wicked, see Peters 2017) problems as holistic, complex, and almost intractable problems, needing cross-disciplinary intersectoral (i.e., transdisciplinary, Boger et al. 2017; Grigorovic et al. 2018) reasoning and solutions. This positions community psychology and CBPR in similar intellectual spaces, whereby Participatory Action Research (PAR), as Fals Borda (1991, 2008) points out, is concerned with co-producing social, cultural, and political change, and rejecting the normative objective research stance for an understanding of social phenomena which is driven by historical learning, current context, and a concern for social justice. It is this politicisation of research which separates community psychology, linked with CBPR, from more established routes to 'doing' psychology in ways which are explicitly more activist in nature. Here, the raising of consciousness of marginalised people, as well as those in positions of power, produced through the research process itself is important. Raising conscious understanding of the power structures which hold people in positions of disadvantage enables the confrontation of, often hidden, inequalities and can result in a legacy of social change not just targeted on current social issues but feeding forward to further instances where the status quo and new orders of oppression can be challenged. These issues of social justice, diversity and equity, *conscientiousization* (Freire 1968), challenging current social orders in order to produce social change and a legacy for social change are explored in the following case study.

The Research Context

The City of Richmond, in British Columbia, Canada is faced with the problem of high housing costs that older, low-income adults struggle to afford. In 2005, members of a housing society, whose primary mandate is the provision of low-cost housing for older adults, raised concerns about the age and quality of one of their housing blocks, a five-acre site built in the 1960s, which consisted of 122 units (24 wooden construction cottages and 98 bachelor suites in a community of 3-story apartment buildings). These units were rented to low-income older adults (aged 60+ years) for independent living at approximately CAN $350 per month. Over time, the safety and liveability in these units became a concern, owing to progressive deterioration of the structures. Indeed, one tenant sustained serious injuries after having fallen through rotten flooring.

By 2010, the housing society partnered with the municipal government to sell two-thirds of their land to a property developer, generating enough capital to redevelop the land on which the low-rises were located. Though tenants of these units were given priority to relocate into the redeveloped housing, they were forced into this situation, many not wanting to move, and they were required to secure their own alternative housing during the rebuild. To alleviate hardship, the housing society helped tenants find new housing, subsidised rents during the construction period, and provided moving support into the new build. Forty of the 152 tenants from the low-rises relocated into the new build. The new build, designed to support tenants' wellbeing and social connectedness, is made up of two 16-story concrete structure towers, totalling 296 one-bedroom units, inclusive of two units designated for two full-time, live-in, multilingual caretakers. Tenants moved into the first tower in March 2015 and the second tower in August 2015.

To ensure that the redevelopment and rehousing of older adults from the low-rises into the new build was completed sensitively and with a remit to involve the voices of older adults, a team of university researchers were invited by the city to be community partners. This partnership was created to: (1) capture sense of place as experienced by the tenants; (2) understand the lived experiences of older adults to inform the provision and programming of effective formal and informal supports within the development; and (3) develop practical guidelines and recommendations for supporting the place-based needs of older adults. This presented an opportunity to use a CBPR approach to document the nuanced meanings of place, identity, and attachment to and detachment from place from the perspective of a sample of low-income, older adults. Moreover, the research sought to (1) highlight issues of diversity and (in)equity concerning the identity and lives of low-income older adults, (2) identify and challenge power relations in the partnership and development process, and (3) challenge existing visions of low-income housing development, certainly concerning how decisions are made and who inputs to those decisions.

Consequently, low-income older adults were positioned as partners in the redevelopment partnership group to enable their concerns about sensitivity to income disadvantage, cultural understandings of their everyday life, support for relocation, and fairness in redistribution of housing to be heard. These issues were extremely important as the older relocated population and hence research participants were a diverse sample of younger and older old adults, 70% Chinese and 30% European, and varied across gender and household composition (with some living alone, others with spouses, and some in larger family groups). The research spanned a three-year period and brought together academics, older adults, city government personnel, property developers, property managers, service providers, and staff from community organisations. The ways in which a community psychology and CBPR worked to challenge the status quo and to raise consciousness of equity and diversity issues in the relocation process, as well as move towards positive social change, are analysed in a case study of forced relocation.

A Case Study of Forced Relocation

The initial problem for researchers involved the creation of space for those who have less societal power to voice their concerns, prioritise important issues, and influence a project which affects them. This meant that researchers worked with tenants pre-move (but post knowledge of relocation) to build trust and ensure they understood their potential role in the production of new housing and had the knowledge and tools to communicate effectively with property developers, property managers, service providers, and other professional stakeholders. There were also huge efforts to manage conflict and to enlighten professionals about the value of older people's involvement. Thus, the team strove to develop an open attitude to the research and to the development both in terms of the style and focus of meetings and by highlighting issues of equity, power, and oppression through the use of participatory methods.

From the outset of the research, a steering committee partnership was created, involving a diverse group of stakeholders who had a vested interest in the wellbeing of tenants and the redevelopment process. Representation from community-based older adults on the steering committee provided the opportunity for their voices to be integrally involved in the project. In addition, community-based research events were hosted, reaching out to the wide range of older tenants of the low-rise apartments and the new build. The events were specifically designed to inform residents about their situation, to gain their perspectives, worries, and disquiet with the relocation processes and to help them explore what they wanted to achieve through the forced relocation process. In terms of community psychology, the researchers positioned themselves as facilitators of voice, while simultaneously raising awareness of the potential for tenant power to positively impact the new build in ways which could promote the development of age-friendly living spaces.

During this initial period, information (i.e., lay summaries) about the research was made available in multiple formats (hard copy and emails) and languages (English and Chinese) so tenants could begin to understand and participate in the research processes. As tenant engagement progressed and more formal points in the research project were reached, data were collected from tenants through a range of participatory methods designed to enable them to articulate their experiences of transitioning from the low-rises to the new build. The research involved the use of in-depth interviews (Legard et al. 2003), photovoice (Wang and Burris 1997) which transformed into a 'photo-tour', community mapping workshops (Fang et al. 2016), a community walk along (Fang et al. 2016) and feedback forums (Fang et al. 2017b). In addition, a series of four deliberative dialogue workshops (Canham et al. 2017; Sixsmith et al. 2017) were conducted with community stakeholders, each building on the interview and photo-tour findings and specifically organised to ensure stakeholder buy-in to the community mapping workshops and the walk along. This constituted an important staged approach (Fang & Place-making with Seniors SFU Research Team 2017) to the research which actualised a plurality of ways of expressing voice and faced all stakeholders with negotiating a network of power relations such that the new build could better fit not just older people's needs, but improve their health, wellbeing and quality of life.

With multilingual researchers who could converse in English, Mandarin, and Cantonese, older English-speaking and Chinese adults were able to share their stories about their lives, their histories, and the requirements of spaces and services to facilitate age-friendly living accommodation. The ways in which tenant participants and steering group

Table 8.1 CBPR data collection methods used in the research

Method	Pre-relocation	Post-relocation	Participants
In-depth interviews	25	–	Tenants
Photo-tours	16	–	Tenants
Deliberative dialogue sessions	*N* = 4: 24 people in each session	–	Professional stakeholders
Feedback forums	–	*N* = 2: 45–50 people in each forum	Tenants
Community mapping workshops	–	*N* = 4: 30–35 people in each workshop	Tenants and professional stakeholders
Community walk along	–	*N* = 1: 25–30 people	Tenants and professional stakeholders

partners were engaged in the research is detailed in Table 8.1, along with trajectory of data collection pre- and post-relocation. The role of community psychology in shaping the partnership and the information garnered to support the new build development are subject to the analysis in Table 8.1.

Steering Committee Meetings (Pre- and Post-relocation) Steering committee meetings were held bimonthly prior to the relocation into the new build, though less frequently post-relocation. Two representatives from the housing society, who were themselves older adults and whose families lived in the old apartments (and subsequently the new build), served as tenant representatives on the committee alongside city personnel and planners, property developers, property managers, services providers, and researchers. The researchers organised and chaired these meetings, provided status updates on the progress of the research, and presented key issues arising from the data collected and engagement with the broad tenant community. By positioning researchers to lead the meetings, meetings were discussion-based rather than simply process and progress oriented. This meant that wider issues such as understanding the heterogeneity of the tenant population and the historical (including personal histories of relocation) and social context of the community were highlighted thus providing important context for decision making. This produced a discursive space which co-produced knowledge across researchers, older people, and the involved professionals, whilst raising awareness of the wider issues of social inclusion faced by tenants.

Despite excitement about the project amongst committee members, the initial meetings were fraught with relationship challenges (Fang & Place-making with Seniors SFU Research Team 2017). For example, some committee members used discriminatory language when talking about culturally diverse groups, particularly persons of Chinese heritage. In one instance, it was suggested that certain cultural groups were wealthy and have displaced white Canadians from their communities. These sentiments created tension with the researchers and those committee members deriving from various ethnic backgrounds. To manage this situation, the research team devised a strategy of acknowledging racist comments but repositioning them within discourse of socially acceptability while redirecting discussions around how to achieve shared goals. This tactic proved successful as problematic language reduced as the research progressed.

Intergenerational challenges were also encountered as some of the researchers and other committee members were young professionals, while the tenant representatives were older. These age and generational differences were expressed in differential understandings of political correctness (e.g., correct terms for identifying ethno-cultural groups) and appropriate behaviours and mannerisms (e.g., ways in which partners spoke and interacted with one another in situations where there are disagreements) during committee meetings. By identifying and discussing such instances, committee members gradually began to share a common framework of communication and respect. This produced a more comfortable space in which the older tenant representatives felt able to contribute effectively to meetings. It was clear that their initial scepticism transformed into respectful requests for information and a sense of joint ownership. Ultimately, the 'them vs. us' attitude became diluted. This is an important consideration for community psychological work, that is, to actively identify and strategize to resolve power imbalances which are codified in language and reintroduce mutually beneficial communication styles.

Another challenge revolved around perceptions of the research goals and the researchers' roles (Fang & Place-making with Seniors SFU Research Team 2017). In terms of community psychology, it is important to ensure that the expectations of all stakeholders are raised and jointly constituted; without that, the chances of unmet expectations can derail chances of achieving positive research outcomes and social change. As the research progressed, it was evident that the tenant representatives developed an unanticipated expectation that the research team would secure and deliver services and programmes in the new build. This conflicted with the research team's expectation that their key role was to enable tenant voice in the process through collecting and reporting on information that could inform service delivery. Again, a direct approach was taken to resolve this situation by discussing different stakeholder roles, clarifying the roles and responsibilities of the researchers during committee meetings. Once the conflicting opinions were detected, they were immediately and directly addressed in an open discussion, enabling all parties to voice their opinions. This method of partnership working resulted in an improved understanding of each partner's role and of the partnership itself. Here, the notion of plurality of views was adopted within the committee and the equal rights of all to challenge those views were promoted.

Difficulties were also experienced when co-creating the research project with the property developers (Fang & Place-making with Seniors SFU Research Team 2017). The disciplinary background and focus of the property developer prioritised the environment in its built form, which, at times, conflicted with the research agenda to focus on understanding and creating supportive community and social environments for older adults (Fang et al. 2018b). The property developers prioritised making money and creating inexpensive housing units because, for them, the build was about creating cheap, affordable housing. The researchers and older tenant representatives challenged the developers' focus on cutting costs. Despite attempting to resolve this issue through individual discussions (between the city and property developers) and in committee meetings (between researchers, the city and property developers), the bottom line revolved around cost cutting. Consequently, post-relocation, tenants complained about the cheap appliances and light bulbs in their units which constantly needed replacing. Replacing light bulbs constituted a danger to tenants who needed to climb steps or stand on a chair to replace them. The property developers failed to grasp that the development of buildings and home environments for older adults requires thoughtful consideration of health and safety issues.

Partnership difficulties between the property managers and the tenant representatives also emerged (Fang & Place-making with Seniors SFU Research Team 2017). The primary goal of the property managers was to successfully occupy the new apartments with tenants who met eligibility criteria: aged 60+ years and low income (i.e., less than CAN $38,000 a year and CAN $250,000 in assets) (Fang et al. 2018b). Having limited experience of working with culturally diverse, low-income older adults on a day-to day-basis, the engagement approach of the property managers often lacked sensitivity to tenant age, ethnicity, and socioeconomic status. For example, during a public forum event, convened as a question and answer session for the tenants, property managers were dismissive of questions, shouted down the tenant's comments, and made little effort to understand the concerns of tenants with limited English language skill (Fang et al. 2017a). To address this challenge, the researchers raised sensitivity awareness issues in committee meetings in an open and respectful manner. Alongside this, findings from the research were presented in meetings. Interestingly, stories of relocation were very powerful in challenging dismissive attitudes or limited knowledge and enabled the older tenant representatives to challenge such attitudes themselves in ways which drew on the findings themselves. This produced an atmosphere of learning and discussion rather than outright conflict.

In spite of these partnership challenges, being part of a transdisciplinary and cross-sectoral partnership (Boger et al. 2017) offered partners a number of opportunities. The older tenant representatives who occupied senior, leadership roles reaped the social benefit of improving housing conditions for tenants. For the housing provider, the partnership resulted in a unique and rewarding opportunity to learn from the research findings about how older adults create a sense of place and achieve a good quality of life in affordable housing units. For the property developers, the partnership offered opportunities to inform how the new build would be received by tenants and how they would use the shared amenity spaces.

Findings from the research enabled a better understanding of how the structural elements in the new build would be used to facilitate social participation amongst tenants. Finally, the property managers gained important skills and knowledge to overcome communication difficulties and build stronger relationships with tenants. By the conclusion of the research, city personnel, the housing provider, property developers, and property managers all expressed appreciation for the perspectives of the tenant representatives and the research findings, which have reportedly helped co-create meaningful housing solutions for older adults who are transitioning into affordable housing.

Building Partnerships Through Participatory Methods: Equity, Power, and Oppression

The committee was not the only vehicle through which the community psychology and CBPR approach integrally impacted the development project. The design and use of a range of participatory methods also contributed strongly to this by feeding into the shared visioning process and enabling shared aims and co-created solutions to emerge. These were: in-depth interviews, photo-tours, deliberative dialogues, community mapping workshops, a community walk along, and feedback forums. The role of community psychology in supporting the development of strong partnership working is described below.

In-depth Interviews Input to Partnership Working (Pre-relocation) Prior to moving into the two new builds, most tenants were eager to talk about their situation because of the stressful nature of the upcoming relocation (Fang et al. 2015). Work with the senior's centre (local to tenants) and talking informally with a range of tenants about the research, its purpose, and transformative potential gave rise to the initial interview schedules and supported an open recruitment strategy. Researchers emphasised the notion of research for future benefit, concern for tenants' situation and 'enabling voice'. The semi-structured, in-depth interviews promoted discussions of participants' housing situation as well as their social positions, identities, and previous transitioning experiences (e.g., from China to Canada) (Fang et al. 2018a). This allowed for a more holistic understanding of participants' relocation experiences, including how they felt within a bureaucratic system which imposed change and how they sought to make sense of and challenge this. Finally, interviews exposed the service and amenity landscape required to support them now and in the future. Interview data were thematically analysed by the research team and presented to the steering committee, ensuring that tenant voices were ever-present during decision-making sessions. Interview findings were critical to partnership building as these data reminded all committee members to remain focused on the tenants (Fang et al. 2015). For example, when aesthetic considerations dominated discussion of the shared amenity space in the new build, tenant requirements for social connectedness were introduced, led by the older representatives, and discussions ensued about how space could be organised to reflect this. Aesthetic notions were then recast into designing the space to encourage social connectedness. More importantly, research findings informed the development of a tenant social committee that organises social activities in the new build. In this way, the research supported current tenants to forward think in terms of a legacy for future social change.

While interviews were an effective data collection tool for some older participants who were reflective and articulate about their lives, other less verbal older participants were enabled to share their experiences of place and relocation through visual means. This ensured a more inclusive voice to be heard by professional stakeholders.

Photo-Tour, Partnership Building, and the Inclusion of Older Adults (Pre-relocation) Capturing visual images of home and community enabled older tenants to transmit meanings through their own data capture (Pink 2013; Mountian et al. 2011). Photovoice is a visual method (Wang and Burris 1997) used to explore personal experiences of a phenomenon while handing the power of voice to participants themselves. Photovoice is usually undertaken by individuals who independently take photographs of situations, people, and places that are meaningful and then select photographs to discuss during an interview. However, several older participants felt uncertain of their ability to collect these data, never having undertaken such a task before. To accommodate these concerns, on discussion with senior centre tenants and researchers, the photovoice technique was amended into a 'photo-tour' whereby the tenant participants led the researchers around their community, selecting photograph locations while simultaneously narrating why specific places held meaning to them. The researchers audio-recorded these narrations in situ and later transcribed these verbatim. The photo-tours not only provided participants with a creative activity but also helped generate important topics of conversation. Participants produced data on their own terms and explained their choices 'in the moment', thus not requiring a later act of memorisation when revisiting the photos to describe them to the researcher. This flexibility of method, driven by participant need, highlights the possibility of a community and participatory approach to think differently about the construction, quality, and value of data, rather than following a preset methodological recipe. For the tenants, photo-tours provided a way to visually portray experiences and discuss personal knowledge about issues that did not emerge in interviews. For example, the importance of pets as family members (see Fig. 8.1) and the ability to play or listen to music (see Fig. 8.2) as ways of maintaining wellbeing and community connection were clearly articulated during photo-tours. The visibility and immediacy of the

Fig. 8.1 The importance of pets for wellbeing and community connection

Fig. 8.2 The importance of music for wellbeing and community connection

photographs occupied a really interesting space in committee meetings and decision-making. On seeing and hearing about the importance of pets, especially for people without family, and the necessity for local social connectedness for those with mobility problems, the design professionals re-thought space use and building mangers changed the new build rules regarding pets. Here, the reality of lives constrained by lack of finances and social engagements were brought into focus and transformed into notions of liveability in the city. As such, notions of the right to a good life for all began to infiltrate collective thinking.

Data co-creation through the photo-tours involved a rich personal analytical process which transformed into the development of a social analytical framework. An unexpected benefit of this technique was the extent to which the storied use of photographs encouraged participants to identify new issues to discuss and foreground aspects of their lives they were proud of, further generating an awareness of their personal agency. However, the difficulties of using this method were also evident as some participants were uncomfortable, anxious, or inhibited in taking photographs.

Deliberative Dialogue as an Inclusive Partnership-Building Mechanism (Pre-relocation) Deliberative dialogue is a method for generating thoughtful discussion, providing an opportunity for concurrently generating and analysing data, engaging participants and synthesising evidence with the end goal of establishing a set of actionable items (Plamondon et al. 2015). As such, the notion of co-producing social change is directly built into the methods and this differs from other public discourse techniques such as debating, negotiating, ideas mapping, and generating consensus (Kingston 2005; Battersby et al. 2017). Deliberative dialogue sessions were designed and organised to create a collective platform for property managers, property developers, local service providers, and representatives from the municipal government to share diverse perspectives and develop potential solutions for creating socially engaging spaces in the new build (see Canham et al. 2017). The dialogue sessions were critical to bringing together professionals who had not previously worked together. The initial aim of the sessions was to develop a shared understanding of the problem area, taking into account different perspectives, reflective of an appreciative inquiry approach (Cooperrider and Whitney 2001). This approach began by exposing and valuing other dialogue participants' expertise and integrating findings from the interviews and photo-tours. Workshop participants then began working towards potential workable solutions. With its focus on change, appreciative inquiry fit well as a deliberative dialogue mechanism as participants proposed ways in which physical, health, leisure, and housing environments can better support tenants in the new build. Deliberative dialogue sessions were transcribed for analysis by the research team.

The focus on exposing and valuing expertise meant that the deliberative dialogue workshops helped to cement partnership relationships as people came to understand gaps in their own knowledge that other professionals could inform. However, difficulties were experienced in supporting professionals (who were secure in their own expertise) to transcend the boundaries of their various knowledge bases, especially when taken-for-granted knowledge was cast into doubt or one's own expertise was perceived to be of lesser value. For example, some participants were perceived as more knowledgeable or powerful than others, which meant that some participants deferred to others, a problem identified by Anyan (2013). With researchers' careful facilitation, active listening, reiteration of the expertise participants held, and reinforcement that all perspectives were equally valued, more trusting and open attitudes developed although it was observed that, in times of disagreement, dominant voices were more listened to, suggesting existing hierarchies and power relations had not been dissolved, rather they had merely been backgrounded.

In terms of producing social change, the dialogues generated possible directions for effective use of shared community spaces in the new build. However, specific actionable solutions were not forthcoming (Canham et al. 2017), perhaps because of the intransigent hierarchies and lack of tenant's voices in the workshops (an intentional decision to equally value diverse professional voices). For instance, a central challenge identified by workshop participants was the lack of financial resources to implement and sustain on-site services and supports. Though solutions were offered (e.g., developing a voluntary tenants' board and fundraising to hire a programme coordinator), no workshop participants made a commitment to solve this challenge. Instead, the onus was placed on the researchers to take action, which was neither feasible nor appropriate. What would have enhanced this deliberative dialogue method is the addition of an 'accountability' feature by which service professionals 'pledge' to carry out an action at the end of the workshop (McCoy and Scully 2002).

Community Mapping (Post-relocation) Participatory mapping is a map-making process that aims to create a visible display of people, places, and experiences (Corbett 2009). This collaborative approach is action oriented and is used to 'enable local people to share, enhance and analyse their knowledge of life and conditions to plan and act' (Chambers 1994, p. 953). Participatory mapping can be used to generate understandings that people construct concerning the resources, services, feelings, and experiences of their local community (Fang et al. 2016). One distinctive aspect of community mapping is that it begins with, prioritises, and values local knowledge, disrupting hierarchical power relations between participants and researchers to attempt to equalise the knowledge space between them (see Wood 2010; Anyan 2013).

Four community mapping workshops were organised as part of the current research (see Fang et al. 2016), creating both a physical and knowledge space where the older tenant participants could interact with professional stakeholder participants (e.g., service providers) and expose meaningful places in the community. At each workshop, four tables were set up, each presenting the same aerial map of the new build and surrounding area (see Fig. 8.3). During the workshops, participants identified locally available places and services, as well as gaps in resources. To support construction of meaning around the map, pens, post-it notes, and flip charts were made available and discussions were digitally recorded and later transcribed. Refreshments were available throughout the workshop.

Differences in participant's perspectives and knowledge meant that communal understandings were difficult to achieve and that the maps created were neither neutral nor unproblematic, as they failed to connect

Fig. 8.3 The community mapping workshop set-up

with frameworks of participant's social positionality and emphasised partiality of knowledge. This meant that participants (usually professional stakeholders) with enhanced local knowledge tended to dominate the sessions. Nonetheless, nuanced understandings of how community spaces work to produce age-friendly communities did emerge. Faced with a new activity that neither tenants nor professional stakeholders had experienced before, mapping partially served as a levelling tool. Here, strong facilitation by the researchers was required to ensure consistency in map-making and positioned them as more dominant voices within the map-making process than was desired. Encouraging workshop participants to walk around the map was an effective way of rebalancing the emerging hierarchical social positions (a silent tenant in one group could speak effectively in another group under different social dynamics).

Further challenges in the mapping process, which prevented tenant participants from fully contributing to the map-making exercise, were identified. Some tenants felt: they could learn nothing new about their community ('We've lived here for more than 20 years, we know everything that is around here'), that their contributions would not be taken seriously or translated into positive change ('I think that no matter what we say, the map will stay the same'), or that their advanced age made their contribution unimportant, given they only have a few years left to live. In addition, many Chinese tenants struggled with translations (translators were present) from English during the rapid talk taking place and noisy venue. The noise was particularly problematic for participants with hearing deficits, effectively silencing them; an issue for consideration in future workshops.

Following all four mapping workshops, participants were invited to complete a postcard where they wrote down one action they could commit to that would improve the ability of tenants to age in place. The aim here was to focus attention on producing action rather than ideas. Some service provider participants felt pressured in their daily work to take action and suggested that the onus should be on tenants to seek out and ask for services. Tenant participants often failed to understand the purpose of the postcards. However, some tenants did make commitments, including commitments to talk to and check on their neighbours more, thereby facilitating social bonds in the new builds.

The Community 'Walk Along' (Post-relocation) As part of the first workshop (Fang et al. 2016), participants were invited to take part in a 'walk along' though only one group did so. Walk alongs have been described as an ethnographic research tool (Kusenbach 2003) designed to generate information through participant's interaction with the environment and are effective in positioning participants as walk leaders (Carpiano 2009; Garcia et al. 2012). This was an effective partnership-building exercise, placing tenants as experts on their community and generating opportunities for companiable (yet revealing) conversations to be had and friendships to emerge (Fang et al. 2018a).

In the current research, the group walk along was used to identify meaningful places, barriers, and facilitators to ageing-in-place (Fang et al. 2016). The walk along process, while difficult (splinter groups formed, noise prevented accurate recordings), generated new ideas to enhance the lives of tenants, particularly as participants were faced with navigating the difficult and deficient landscape in the surrounding community. Critically, these navigation challenges were witnessed first-hand by professional stakeholder participants who subsequently sought to design solutions to them. For example, on observing difficulties in crossing a major road to reach local shops from the new build, the city installed a new traffic control system that allows tenants more time and dedicated space to safely cross the street and directly access the shopping mall, thereby enhancing their social participation (Fig. 8.4).

Group walk along challenges were evident: disagreements about which route to take, accommodating less mobile participants, and ensuring the safety of splinter groups. Researchers acted to balance the need for direction, maintaining safety and comfort while giving participants full control. As the routes were emergent rather than pre-determined, such issues need to be navigated in situ. Despite best efforts, the data gained were tempered by researcher control and participant (dis)comfort. Scouting out the community prior to the walk along would have been useful in ensuring adequate resting places were available and safety (especially when crossing roads) was prioritised.

Fig. 8.4 The community walk along the main road

Feedback Forums and Tenants' Inclusion in the Research (Post-relocation) To determine what tenants thought about living in the new build, the participatory research process, and their inclusion in this process, two feedback forums were held in the new build's amenity space. Prior to these events, a sustainability and implementation plan produced by the research team (Fang et al. 2017) was emailed to tenants along with a summary of findings and implementation solutions based on the research (Fang et al. 2015). In addition, promotional materials were designed in English and Chinese languages and distributed throughout the new build. In a poignant example of the need to partner with those whose lives you are aiming to influence, these promotional materials had included what the researchers thought was the simplified Chinese character for 'Towers' that was read by the Chinese tenants as 'Urns'. Upset at the implication that they were living in an urn, tenants contacted the housing society whose director organised with the researchers to amend the promotional materials.

During the feedback forums, tenants were provided with refreshments and information about the research findings (both verbally and via printed appropriate language lay summaries). The forums, open to all tenants in the new build, were organised around two discussion tables, one for Mandarin/Cantonese speakers and the other for English speakers. At each table, the discussions began with the very general question: 'How do you feel about the new build?' Discussions, guided by tenants, lasted 60–90 minutes, were recorded and transcribed.

The forums served several purposes: property managers were keen for the events to occur so tenants would view them as caring landlords, researchers used them to collect data, and tenants gained information and a chance to socialise with others in the new build. In terms of the community psychology approach adopted, the feedback forums worked well to finalise the partnership with a strong focus on what had been achieved and from all participant contributions. Researchers simply leaving the field once data were collected were never part of the research plan, and the ethos of partnership was carried through to the end.

Concluding Remarks

Developing a safe, secure, and socially engaging living environment for older adults to age in the right place (Golant 2015) requires careful consideration of how individuals make meaning, identify, connect, and interact with the physical environment (Fang et al. 2018a). The project began with a conviction that older tenant and other stakeholder voices need to be equally heard and that this should lead to the co-creation of a shared vision, aim, and action plan, and co-production of workable solutions which strive to produce positive social change. In this case, social change focused on a new age-friendly apartment building in a well-serviced community, accessible and enjoyable for low-income older adults. As such, it was important to acknowledge and understand the lives of older people and firmly establish their inclusion in the project at the outset, as well as during planning and development phases (Woolrych and Sixsmith 2013).

This proved to be a challenge as the professional stakeholders involved often lacked an understanding of the needs of older people and were more concerned about the physical aspects of the building and the costs (Fang et al. 2018b). Nevertheless, a collaborative, team-working community psychology approach was constructed which helped to raised awareness of social cultural and political key issues, indeed to focus on conscientiousization beyond costs and physical environment such as ageing in a well-resourced community, equity in terms of living a good life, and designing for cultural and ethnic diversity. Construction of this approach was not pre-determined at the project's outset; rather the principles and values of community psychology were set at outset and transmuted into project actions as the project moved forward. Here, the researchers needed to be extremely flexible, to identify problems early before they became entrenched, and to work solutions together with all stakeholders while privileging the voice of the older tenants. Perhaps as important was the emphasis on (1) *working together* across disciplinary, expertise, and sectoral boundaries (rather than just talking about issues); (2) *moving together* through each other's space (as in the mapping workshops and walk alongs); (3) focusing specifically on *learning to listen* to each other and *value expertise* (via interview and photo data and deliberative dialogues); and (4) adopting a *just solution and action-oriented* perspective (in mapping workshops and deliberative dialogues).

This collaborative way of working applied a community ethos to the partnership committee as well as the multiple participatory research methods which provided older adults, community partners, and local professional stakeholders with various opportunities to contribute to decision-making and enabled the articulation of multiple viewpoints throughout the redevelopment process. This happened, simultaneously, as researchers relayed research findings from tenants to the steering committee, as well as through direct discussions between tenants and professional stakeholders.

Despite best efforts, aspects of the approach were difficult to negotiate and while successes were achieved, several tensions of working across stakeholder voices and co-producing viable solutions were encountered. These pointed to the need to (1) pay attention to the complex social relationships which bind people in partnerships, (2) use multiple methods to capture as holistic a picture of complex community phenomenon as possible, and (3) the integral involvement of older adults in the decision-making and knowledge generation process. Working in community psychological ways does not constitute a one-way battle to reveal and instantiate the voice of marginalised people such as low-income older adults but depends on mediation, negotiation, and a will to jointly co-create socially just solutions to wicked problems.

Earlier research has suggested that participatory working and techniques such as community mapping are useful tools to encourage collaboration as well as dialogue and relationship building among research participants (Amsden and VanWynsberghe 2005). In the current case, this was extremely important given that the requirement to move home was forced upon older tenants with little personal and financial resources. Compounding such difficulties, cultural, ethnic, historical, and language problems pervaded the research and development space. In these circumstances, involving older adults requires careful preparation. Such preparation involves:

- understanding by researchers and other professional partners of the valuable experiential expertise that older adults bring to research;
- mediation and negotiation during steering committee meetings and research events to avoid confrontational situations and develop an ethos of shared aims, respect, trust, and openness;
- adopting a cultural and ethnic lens;
- accommodation of research participants' sensory and mobility limitations;
- attention to social aspects of research events and partnership processes; and
- planning for different forms of data collection to gather a diverse set of views.

One of the key messages deriving from this project relates to the requirement of researchers to negotiate power plays and power differentials which emerged at different points in the process. Identifying and discussing such dynamics as they occurred was an important step in the co-creation and co-production process. The research also highlighted the need to adapt methods to fit the requirements not just of the research questions but also of the research participants and the dynamics of the partnership concerned (Fang & Place-making with Seniors SFU Research Team 2017). Finally, if such projects are to deliver equitable solutions and produce positive social change, then there needs to be a focus on developing and implementing solutions with an explicit action-oriented and social justice stance (Woolrych 2017).

This chapter has detailed the nuances of involving older Chinese and English-speaking tenants in partnership working during a research process based in the principles and values of community psychology. The research confirms that while partnering with older adults can be complicated, it is necessary. By striving to do the best possible research, within a constrained relational landscape, some good, if not perfect, solutions can be produced. The older tenants in this project ended up with a viable solution to living well in their city, one in which home, apartment block, and community were equally considered alongside a series of competing stakeholder perspectives. The lessons learnt were considerable: older adults learnt much about the research and the development process, professional stakeholders learnt much about the value of including older people in the project, the researchers learnt much about doing research in difficult relational contexts, but that it is only by working together *with* older adults can we best understand how to support them in living independently and ageing well in the right place to sustain their health, wellbeing, and quality of life.

References

Amsden, J., & VanWynsberghe, R. (2005). Community Mapping as a Research Tool with Youth. *Action Research, 3*(4), 357–381.

Anyan, F. (2013). The Influence of Power Shifts in Data Collection and Analysis Stages: A Focus on Qualitative Research Interview. *The Qualitative Report, 18*(18), 1–9.

Battersby, M. L., Fang, M., Canham, S., Sixsmith, J., Moreno, S., & Sixsmith, A. (2017). Co-creation Methods: Informing Technology Solutions for Older Adults. In Z. Jia & G. Salvendy (Eds.), *Human Aspects of IT for the Aged Population: Aging, Design and User Experience* (pp. 77–89). New York: Springer.

Boger, J., Jackson, P., Mulvenna, M., Sixsmith, J., Sixsmith, A., Mihailidis, A., et al. (2017). Principles for Fostering the Transdisciplinary Development of Assistive Technologies. *Disability and Rehabilitation: Assistive Technology, 12*(5), 480–490.

Buffel, T., Phillipson, C., & Scharf, T. (2012). Ageing in Urban Environments: Developing 'Age-Friendly' Cities. *Critical Social Policy, 32*(4), 597–617.

Campbell, C., & Murray, M. (2004). Community Health Psychology: Promoting Analysis and Action for Social Change. *Journal of Health Psychology, 9*(2), 187–195.

Canham, S. L., Fang, M. L., Battersby, L., Woolrych, R., Sixsmith, J., Ren, T. H., & Sixsmith, A. (2017, e-pub). Contextual Factors for Aging Well: Creating Socially Engaging Spaces Through the Use of Deliberative Dialogues. *The Gerontologist*. https://doi.org/10.1093/geront/gnx121

Carpiano, R. M. (2009). Come Take a Walk With Me: The "Go-along" Interview as a Novel Method for Studying the Implications of Place for Health and Well-being. *Health Place, 15*(1), 263–272.

Chambers, R. (1994). The Origins and Practice of Participatory Rural Appraisal. *World Development, 22*(7), 953–969.

Cooperrider, D. L., & Whitney, D. (2001). A Positive Revolution in Change: Appreciative Inquiry. *Public Administration and Public Policy, 87*, 611–630.

Corbett, J. (2009). Good Practices in Participatory Mapping: A Review Prepared for the International Fund for Agricultural Development (IFAD). Rome: International Fund for Agricultural Development

Davies, J. K., & Kelly, M. (2014). *Healthy Cities: Research and Practice*. Abingdon: Routledge.

Devine-Wright, P., & Lyons, E. (1997). Remembering Pasts and Representing Places: The Construction of National Identities in Ireland. *Journal of Environmental Psychology, 17*, 33–45.

Dixon, J., & Durrheim, K. (2000). Displacing Place-Identity: A Discursive Approach to Locating Self and Other. *British Journal of Social Psychology, 39*, 27–44.

Fals Borda, O. (1991). Some Basic Ingredients. In O. Fals Borda & A. Rahman (Eds.), *Action and Knowledge: Breaking the Mono- Poly with Participatory Action Research* (pp. 3–12). New York: Apex Press.

Fals Borda, O. (2008). The Application of the Social Sciences' con- Temporary Issues to Work on Participatory Action Research. *Human Organization, 67*, 359.

Fang, M. L., & Place-making with Seniors SFU Research Team. (2017a). *Place-making with Seniors: Towards Meaningful Affordable Housing: Final Report*. Vancouver: Gerontology Research Centre.

Fang, M. L., Woolrych, R., Sixsmith, J., Canham, S., Battersby, L., & Sixsmith, A. (2015). *Our Place, Our Space: Resident Stories and Place Priorities*. Available: http://www.sfu.ca/content/dam/sfu/starinstitute/Reports/English Sense of Place.pdf

Fang, M. L., Woolrych, R., Sixsmith, J., Canham, S., Battersby, L., & Sixsmith, A. (2016). Place-making with Older Persons: Establishing Sense-of-place Through Participatory Community Mapping Workshops. *Social Science & Medicine, 168*, 223–229.

Fang, M. L., Battersby, L., Canham, S., Ren, T. H., Woolrych, R., Sixsmith, J., & Sixsmith, A. (2017b). *Aging Well at Home: An Implementation and Sustainability Plan*. Available: http://www.sfu.ca/content/dam/sfu/starinstitute/Reports/Implementation Sustainability Guidelines_28April2017_Final.pdf

Fang, M. L., Woolrych, R., Sixsmith, J., Canham, S. L., Battersby, L., Ren, T. H., & Sixsmith, A. (2018a). Integrating Sense-of-place Within New Housing Developments: A Community-based Participatory Research Approach. In A. M. Goulding, S. B. Davenport, & A. Newman (Eds.), *Creative Practice in the Resilience of Older People*. Bristol: Policy Press.

Fang, M. L., Canham, S., Sixsmith, J., Woolrych, R., & Sixsmith, A. (2018b). Designing Homes for Seniors: The 'Ins' and 'Outs' of an Affordable Housing Project in Western Canada. *Design Community, 4*(86), 72–85.

Freire, P. (1968). *Pedagogy of the Oppressed*. New York: Continuum.

Garcia, C. M., Eisenberg, M. E., Frerich, E. A., Lechner, K. E., & Lust, K. (2012). Conducting go-along Interviews to Understand Context and Promote Health. *Qualitative Health Research, 22*(10), 1395–1403.

Golant, S. M. (2015). *Ageing in the Right Place*. Baltimore: Health Professions Press.

Grigorovich, A., Fang, M., Sixsmith, J., & Kontos, P. (2018). Defining and Evaluating the Effectiveness of Transdisciplinary Research in Aging and Technology. *BMC Geriatrics*.

HAPPI. (2012). *Housing Our Ageing Population: Plan for Implementation*. https://www.housinglin.org.uk/_assets/Resources/Housing/Support_materials/Other_reports_and_guidance/Housing_our_Ageing_Population_Plan_for_Implementation.pdf. Accessed 5 Dec 2018.

Hwang, E., Cummings, L., Sixsmith, A., & Sixsmith, J. (2011). Impacts of Housing Adaptations on Aging-in-place. *Journal of Housing for the Elderly, 25*(3), 246–257.

IDGO. (2012). *Inclusive Design for Getting Outdoors*. https://www.idgo.ac.uk/pdf/Intro-leaflet-2012-FINAL-MC.pdf. Accessed 5 Dec 2019.

Kagan, C., Burton, M., Duckett, P., Lawthom, R., & Siddiquee, A. (2011). *Critical Community Psychology* (1st ed.). West Sussex: Wiley-Blackwell.
Kingston, R. J. (2005). *Public Thought and Foreign Policy: Essays on Public Deliberations About Americans' Role in the World* (1st ed.). Washington, DC: Kettering Foundation Press.
Kusenbach, M. (2003). Street Phenomenology: The Go-along as Ethnographic Research Tool. *Ethnography, 4*(3), 455–485.
Legard, R., Keegan, J., & Ward, K. (2003). In-depth Interviews. In J. Ritchie & J. Lewis (Eds.), *Qualitative Research Practice: A Guide for Social Science Students and Researchers* (pp. 138–169). London: Sage.
Lewicka, M. (2013). Localism and Activity as Two Dimensions of People–place Bonding: The Role of Cultural Capital. *Journal of Environmental Psychology, 36*, 43–53.
Manzo, L. C., & Perkins, D. D. (2006). Finding Common Ground: The Importance of Place Attachment to Community Participation and Planning. *CPL Bibliography, 20*(4), 335–350.
McCoy, M. L., & Scully, P. L. (2002). Deliberative Dialogue to Expand Civic Engagement: What Kind of Talk Does Democracy Need? *National Civic Review, 91*(2), 117–135.
Mountian, I., Lawthom, R., Kellock, A., Duggan, K., Sixsmith, J., Kagan, C., & Purcell, C. (2011). On Utilising a Visual Methodology: Shared Reflections and Tensions. In P. Reavey (Ed.), *Visual Methods in Psychology: Using and Interpreting Images in Qualitative Research* (pp. 346–360). London: Routledge.
Oswald, F., & Wahl, H. W. (2003). Place Attachment Across the Life Span. In J. R. Miller, R. M. Lerner, L. B. Schiamberg, & P. M. Anderson (Eds.), *The Encyclopedia of Human Ecology* (pp. 568–572). Santa Barbara: ABC-CLIO Inc.
Peters, B. G. (2017). What Is So Wicked About Wicked Problems? A Conceptual Analysis and a Research Program. *Policy and Society, 36*(3), 385–396.
Pink, S. (2013). *Doing Visual Ethnography*. London: SAGE.
Plamondon, K. M., Bottorff, J. L., & Cole, D. C. (2015). Analyzing Data Generated Through Deliberative Dialogue: Bringing Knowledge Translation Into Qualitative Analysis. *Qualitative Health Research, 25*(11), 1529–1539.
Proshansky, H. M., Fabian, A. K., & Kaminoff, R. (1983). Place-identity: Physical World Socialization of the Self. *Journal of Environmental Psychology, 3*(1), 57–83.
Relph, E. (1976). *Place and Placelessness*. London: Pion.

Scharf, T., Phillipson, C., & Smith, A. (2005). Social Exclusion of Older People in Deprived Urban Communities of England. *European Journal of ageing, 2*(2), 76–87.

Seamon, D. (2014). Place Attachment and Phenomenology: The Synergistic Dynamism of Place. In L. C. Manzo & P. Devine-Wrights (Eds.), *Place Attachment: Advances in Theory, Methods and Applications* (pp. 11–22). New York: Routledge.

Sixsmith, J. (1986). The Meaning of Home: An Exploratory Study of Environmental Experience. *Journal of Environmental Psychology, 6*, 281–298.

Sixsmith, A., & Sixsmith, J. (2008). Ageing in Place in the United Kingdom. *Ageing International, 32*(3), 219–235.

Sixsmith, J., Fang, M. L., Woolrych, R., Canham, S., Battersby, L., & Sixsmith, A. (2017). Ageing Well in the Right Place: Partnership Working with Older People. *Working with Older People, 21*(1), 40–48.

Sixsmith, J., Fang, M., Woolrych, R., Canham, S., Battersby, L., & Sixsmith, A. (2018). Understanding Sense-of-place for Low-Income Seniors: Living at the Intersections of Multiple Oppressions, Positionalities and Identities. In O. Hankivsky & J. Jordan-Zachery (Eds.), *Bringing Intersectionality to Public Policy*. London: Palgrave Macmillan.

Tuan, Y. F. (1977). *Space and Place: The Perspective of Experience*. Minneapolis: University of Minnesota Press.

Twigger-Ross, C. L., & Uzzell, D. L. (1996). Place and Identity Processes. *Journal of Environmental Psychology, 16*, 205–220.

Wang, C., & Burris, M. A. (1997). Photovoice: Concept, Methodology, and Use for Participatory Needs Assessment. *Health Education and Behavior, 24*(3), 369–387.

WHO. (2007). *Global Age-friendly Cities: A Guide*. Geneva: WHO. http://www.who.int/ageing/publications/Global_age_friendly_cities_Guide_English.pdf. Accessed 5 Dec 2018.

Wood, D. (2010). *Rethinking the Power of Maps*. New York: Guilford Press.

Woolrych, R. (2017). Delivering Age-Friendly Environments: Social Justice and Rights to the City. *The Planner, 169*, 10–11.

Woolrych, R., & Sixsmith, J. (2013). Mobilising Community Participation and Engagement: The Perspective of Regeneration Professionals. *Journal of Urban Regeneration and Renewal, 6*(3), 309–321.

Woolrych, R., & Sixsmith, J. (2018). Place, Space and Displacement: Gentrification and Urban Regeneration in the UK. In A. Portella & G. Pereira (Eds.), *Insights on Favelas*. Pelotas: Universidade Federal de Pelotas. UFPEL E-books.

9

Community Connections and Independence in Later Life

Charles Musselwhite

Introduction

Traditionally transport for older people has been considered an abstract concept that has not fully taken into account the social context. Clearly, issues of ageing, mobility, and transport and the relationships between these are important. Yet the traditional decontextualised approach to transport policy and practice has had less than ideal consequences for society in general, and older people in particular. As a result, the system has been dominated by private motor vehicles at the expense of the environment and personal health and safety, creating a society dependent on oil; a society with severed residential areas with associated eradication of local services, shops, and provision; and an unhealthy acceptance of injury, illness, and death. Ignoring the social element of transport has reduced the concept of travel and transport to a mere mechanism of

C. Musselwhite (✉)
Centre for Innovative Ageing, Swansea University, Swansea, UK
e-mail: c.b.a.musselwhite@swansea.ac.uk

E. Peel et al. (eds.), *Psychologies of Ageing*,
https://doi.org/10.1007/978-3-319-97034-9_9

getting to a destination as quickly and efficiently as possible for the majority, at the exclusion of localness and the positive utility of the journey itself.

The growing disciplines of traffic and transport psychology, the mobilities movement in sociology, and the cultural spaces and mobility movement in human geography emphasise the importance of placing people at the centre of investigating and understanding transport and mobility. Such disciplines stress the importance of examining how individuals shape and are shaped by transport and how they embody and experience transport in relation to culture. There is a growing acknowledgement of such approaches in policy and practice, along with a change in focus from delivering efficient, speedy, and safe mobility to one about facilitating movement of differing speeds, encompassing motivations and balancing needs of movement with those impacted upon by this movement. This change has been brought about in some respects by changes in cultural and political philosophy and ideology. State focus is on encouraging individuals to change their own behaviour, as well as a focus on behavioural change and nudging (see Avineri and Goodwin 2010, for a review) rather than enforcing or regulating change.

Hence, a shift towards understanding that the social elements intertwined with transport use is vital for a true realisation of transport studies. Haglund and Aberg (2000) argue that traffic and transport should be viewed as a social situation where drivers interact and influence each other. O'Connell (2002) notes that transport studies "must not be based on an erroneous model of humans as abstract rational actors, isolated from their social context and operating on purely 'objective' criteria" (p. 201). Thus this chapter introduces ecological framework models coupled with critical gerontological theory to explain older people's mobility and uses the model to explain how to improve mobility for older people for those who have given up driving.

Models to Explain Links Between Social Environments and Transport and Mobility

Ecological models are increasingly being used to explain the interactional relationship between the social and physical context and an individual's behaviour. Examples of ecological models used in terms of ageing and the transport and mobility context include Webber et al.'s (2010) conical model of mobility and Bronfenbrenner's ecological systems model (Brofenbrenner 1979, 1989, 2005) which have been applied to transport (Musselwhite et al. 2014; Musselwhite 2016; Ormerod et al. 2015). Webber's model is based on different layers of environment from home to the world and view mobility through five determinants (cognitive, psychosocial, physical, environmental, and financial), with gender, culture, and biography (personal life history) viewed as cross-cutting influences. Webber et al. (2010) view mobility as the ability to move oneself (e.g., by walking, by using assistive devices, or by using transportation) within community environments from home, to neighbourhood, and regions beyond. Each of the five determinants interacts and varies from day to day within individuals and between individuals.

Originally, Bronfenbrenner's model was applied to child development, explicitly addressing children's play and the wider environment (Holt et al. 2008) and work-life balance in families (Kulik and Rayyan 2006). It has more recently been used in environmental gerontology to address rural ageing (Keating and Phillips 2008) and risk taking and transport (Musselwhite et al. 2014) as well as older people's mobility (Musselwhite 2016; Ormerod et al. 2015). Bronfenbrenner's ecological model (see Fig. 9.1) proposes five layers: the microsystem, the mesosystem, the exosystem, the macrosystem, and the chronosystem that circle concentrically around the individual. The *microsystem* is the layer closest to the individual, consisting of an environment with which the individual immediately interacts, for example the social environment of immediate family and peers and physical and natural environments including house, buildings, roads, and greenery in the immediate neighbourhood. It includes both the actual objective elements found in the environment but also, crucially, the subjective experience of interacting with these elements, for example elements that make the house a home (Keating and Phillips 2008; Lawton 1999; Peace et al. 2006). The *mesosystem* layer provides the connection between the different structures of the microsystem, for example how the social and built environment interact and interlink (e.g., Berk 2000). The *exosystem* layer encompasses the wider social system within which the individual does not function directly but impacts through the mesosystem and microsystem. This level includes people's connections to significant others including wider networks of family

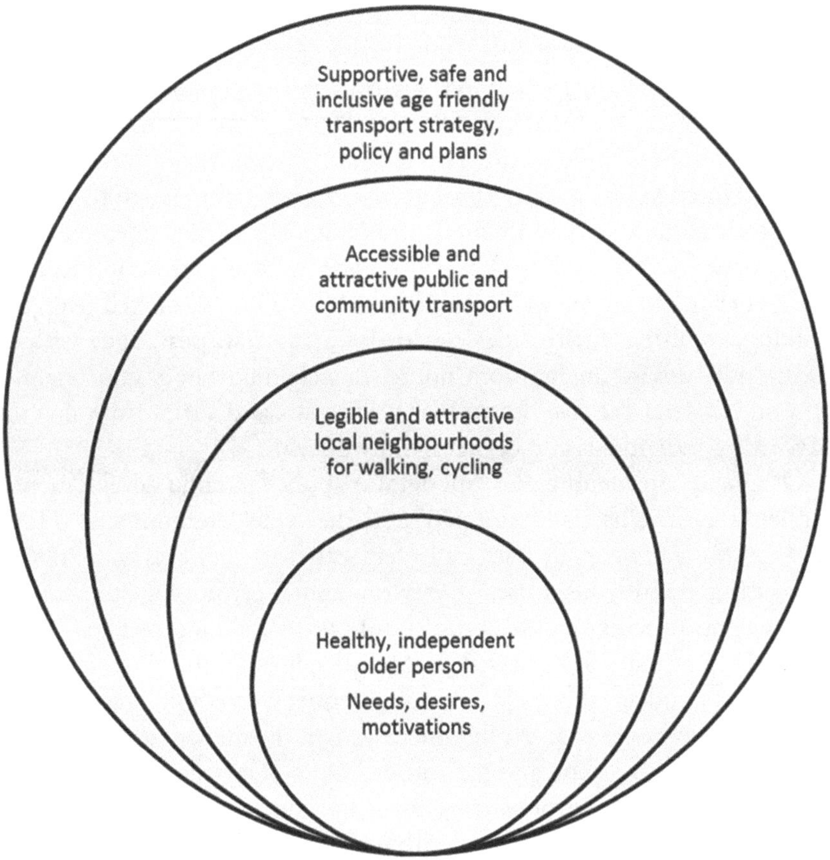

Fig. 9.1 Domains of an age-friendly transport system (after Musselwhite 2016)

members, friends, and neighbours who can be a source of social capital that helps people interact with and navigate their surroundings (Keating and Phillips 2008; Peace et al. 2006). This also includes elements such as the policy environment and laws, programmes, and regulations (Lawton 1999). The outermost layer, the *macrosystem,* comprises societal elements such as culture, values, customs, and laws (Berk 2000). In addition to these four layers, a *chronosystem* adds a dimension of time, for example, the physiological changes that occur with ageing.

To this perspective can be added a layer of critical thinking which emphasises the importance of the model being bi-dimensional, meaning the individual is not simply a victim of the environment but one who acts on the world. The resulting critical ecological model has been linked previously with critical gerontology in examining rural behaviour for older people (Keating and Phillips 2008) and transport and mobility of older people (Ormerod et al. 2015). The critical approach to gerontology stems from concerns with inequalities in power within society which marginalise older people (Phillipson and Walker 1986). More recently this concern has been applied to a spatial approach and environment context, examining the relationship between the imposed accepted wisdoms of disengagement with the environment in later life and the agency, adaptation, and reconstructive abilities of older people to shape their environment (Phillipson et al. 2000; Peace et al. 2006). As noted in Ormerod et al. (2015), spaces, places, and the links between them can exclude older people through the over-reliance on, for example, private mobility, the preserve of the young and fit, at the expense of public, community, or even active forms of mobility (Keating and Phillips 2008). At the same time, a critical ecological perspective notes that there is a multitude of experiences of ageing which differ in different contexts and different places and space (Ormerod et al. 2015) and that any findings must do justice to this diversity.

The "Hypermobile World"

Transport is more important to older people than ever before. In most wealthy and many low- and middle-income countries, high levels of mobility are traversed in order to stay connected to communities, friends, and family and to access shops and services. The car has been central to this hyper-connectivity, affording more choice over the location of work and home. Society has become so organised around the car, that those without a vehicle can become socially excluded. Such institutional car-ism occurs without acknowledging the wider negative consequences of priority planning around the car, excluding non-car users, pollution, severance of communities and accidents, and associated casualties.

Not surprisingly in many Western cultures the growth of car use has been across the life course and there has been a huge increase in the number of older drivers on the road. At the same time walking, cycling, and non-urban bus use have been in decline across all ages. This has significant consequences for older people when they have to give up driving and use other modes of transport, which they may not have used for many years. The decline in use of such services, especially bus services, means they may be infrequent in nature and hence difficult to use.

Being mobile in later life is linked to quality of life (Schlag et al. 1996). Giving up driving, in particular, is linked to a decrease in wellbeing and an increase in mental and physical health problems. This is due to both a reduction in ability to get out and about but also related to psychological issues associated with freedom, status, norms, and independence (Edwards et al. 2009; Fonda et al. 2001; Ling and Mannion 1995; Marottoli et al. 1997, 2000; Mezuk and Rebok 2008; Musselwhite and Haddad 2010; Musselwhite and Shergold 2013; Peel et al. 2002; Ragland et al. 2005; Windsor et al. 2007; Ziegler and Schwannen 2011). Older men find it more difficult to give up driving and are more likely to have to be told to give up driving (Musselwhite and Shergold 2013). This appears to be due to two reasons: (1) males have a greater more emotional bond with their vehicles than females, it is often linked more closely to self-identity and self-image, relating to sense of purpose, youthfulness, and pride (Musselwhite and Shergold 2013), and (2) females spend more time planning to give up driving than men, including time getting used to other forms of transport (Musselwhite and Shergold 2013).

It is vital to create a transport system for older people, so they do not have to suffer health and wellbeing issues when they give up driving. Musselwhite (2016) proposed an age-friendly transport system built around Bronfenbrenner's ecological model of human behaviour, starting with the person at the centre and working out to laws, policy, and plans at the outside, with neighbourhood and public and community transport provision in between (Fig. 9.1). The main points of this model are elaborated later in relation to psychology.

Point 1: Transport System That Is Planned Around the Needs, Desires, and Motivations of Older People's Mobility

Musselwhite and Haddad (2010) propose a model of needs and motivations for travel in later life around three main levels of need (Fig. 9.2). The levels are hierarchical based on how aware participants are of the need. Musselwhite and Haddad (2010) used reconvened focus groups and interviews with the same older participants to develop this model. The needs mentioned first, at the base of the table, are the practical or utilitarian needs—the need to get from A to B as quickly, reliably, safely, and cheaply as possible. The next level is the psychosocial needs, related to how travel fulfils a need for independence, control, and the need to be seen as normal in society and how this relates to roles, identity, self-esteem, and impression management. The top level of need, articulated much later on by participants in the research, was the need to travel for its own sake, to get out and about, to people watch, to see nature, to test their own ability. This top level was termed as aesthetic needs. Travel or mobility in later life is important at all three levels, yet they have differing ways of being presented socially. It is very common for older people, practitioners, and policymakers to talk about the need for travel at the

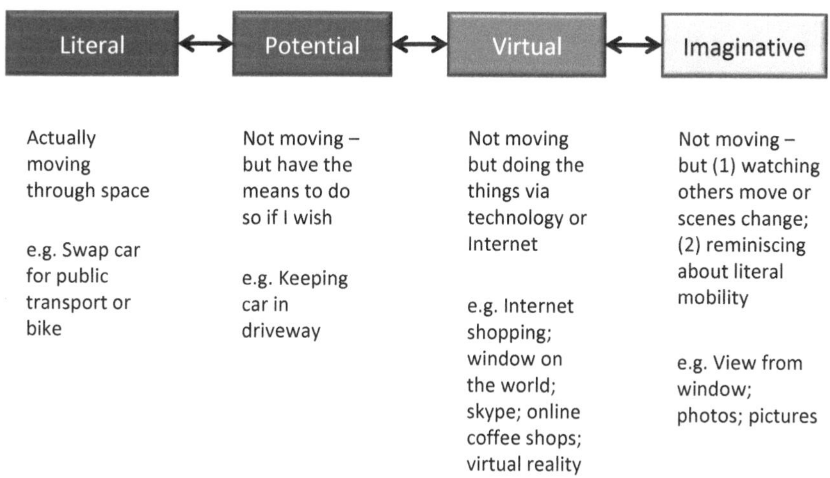

Fig. 9.2 Hierarchy of travel needs in later life (after Musselwhite and Haddad 2010)

utilitarian level and less common for them to discuss the social or affective needs. Even less common is the discussion of travel for its own sake or for "luxury" or "discretionary" purposes; the aesthetic needs. Hence, transport provision in later life is usually centred on practical or utilitarian support while forgetting other important levels of need. So older people with mobility difficulties who may have given up driving can actually get their utilitarian needs satisfied somewhat (though this can still be difficult) by, say, community transport, but their social, affective, and aesthetic needs are not met. In our hypermobile world, driving a car readily fulfils all three level of needs.

Psychosocial needs for driving have been highlighted in previous research. Mollenkopf et al. (2011) identified seven needs that older participants stated were important for them and had remained important for ten years. These were out-of-home mobility as a basic emotional experience; physical movement as a basic human need; mobility as movement and participation in the natural environment; mobility as a social need; mobility as an expression of personal autonomy and freedom; mobility as a source of stimulation and diversion; and finally the ability to move about as a reflective expression of the person's remaining life force. This psychosocial element is especially absent if driving is stopped. As Musselwhite and Haddad (2010) and Zeigler and Schwannen (2012) note those who stopped driving felt a particular loss of independence, especially in very car-dependent areas and for those who had been frequent drivers throughout their life (Adler and Rottunda 2006; Davey 2007; Siren and Hakamies-Blomqvist 2009).

Parkhurst et al. (2014) found that older people's mobility may involve other forms than simply physical or literal and proposed a model that involves virtual, potential, and imaginary mobility as well (see Fig. 9.3). Virtual mobility refers to the use of computing and information technology to satisfy mobility needs, for example keeping in touch with family and friends via telephone or Skype, ordering and having shopping delivered online, and the advent of telehealth. Potential mobility, first coined by David Metz (2000), is the perceived ability to be able to go anywhere, when and how often an individual wants (Metz 2000). Musselwhite and Haddad (2010) note how the car fulfils the potential for travel perfectly,

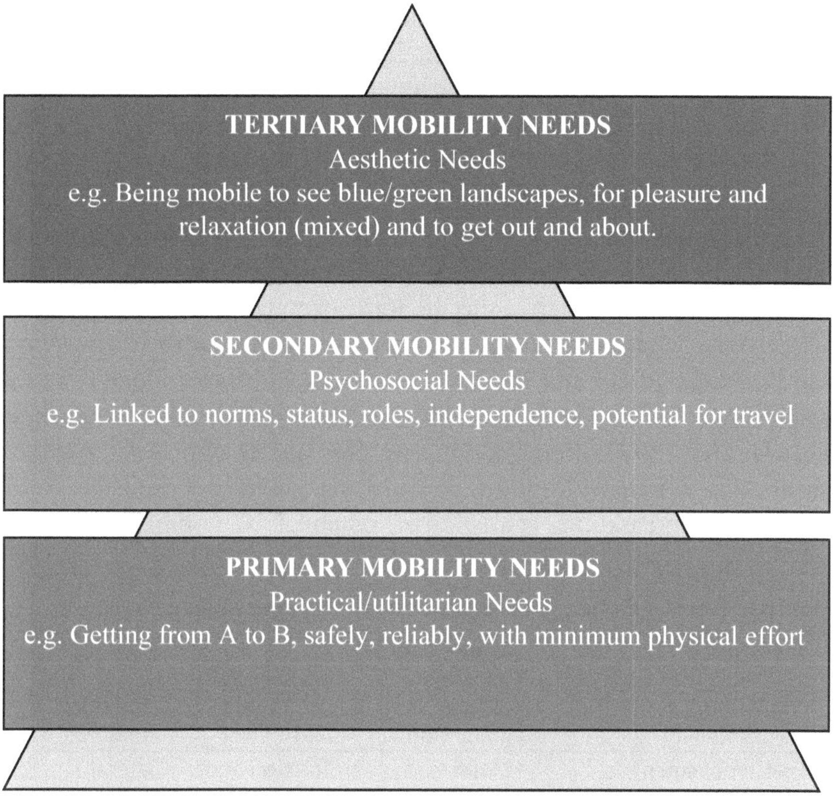

Fig. 9.3 A continuum of modes for connectivity (after Parkhurst et al. 2014)

highlighting examples of older people unwilling to give up the car even when they barely use it, just in case they may need it for various trips. Shergold et al. (2012) term this potential as motility and include the perceived ability to use different modes, for example knowing and understanding the norms of a bus adds to the ability to use it, not just having a bus service outside. Imaginary mobility encompasses two strands: first how literal travel is replaced by construction of travel and mobility in the mind, communicated through storytelling or art, maybe based on previous experiences. Second, imaginary mobility might refer simply to observing movement from a still place, such as looking out of the window (see also Musselwhite 2014) or watching television.

Ziegler and Schwanen (2011) provide a similar taxonomy in their study of older people in County Durham. They proposed imaginary and electronic mobility as being important in later life which equate similarly to the virtual and imaginative mobility proposed by Parkhurst et al. (2014). Mobility practices equate with literal mobility showing the importance of literal and temporal practices to maintain daily life. Mobility of the self is the need, motivation, or disposition to connect with the world and is similar to potential for travel but is linked more towards a will to remain connected socially than to a specific form of transport. Table 9.1 maps Ziegler and Schwanen's (2011) elements of mobility with Musselwhite and Haddad's (2010) and Parkhurst et al.'s (2014) dimensions. Mobility practices are clearly related to practical and utilitarian needs in Musselwhite and Haddad's (2010) model and in turn this equates with literal mobility in Parkhurst et al. (2014) model. The need to connect to the outside world maps well with Parkhurst et al.'s (2014) potential for mobility and is part

Table 9.1 Types of mobility in later life comparing three typologies

Type of mobility	Ziegler and Schwanen (2011) dimension	Musselwhite and Haddad (2010)	Parkhurst et al. (2014)
Physical movement	Mobility practices	Practical and utilitarian needs	Literal mobility
The need to connect to the outside world	Mobility of the self	Psychosocial and affective needs	Potential mobility
Attitudes towards mobility practices	Attitudes towards mobility practices	Psychosocial and affective needs	All levels of mobility
Memory and imagination that recollect or construct events	Imaginary mobility	Aesthetic needs	Imaginative mobility
Using electronic communication, such as the Internet, to substitute for mobility	Electronic mobility	All levels of need	Virtual mobility

of Musselwhite and Haddad's (2010) psychosocial needs. Attitudes towards mobility practices relate to Musselwhite and Haddad's (2010) psychosocial needs but are not clearly related to any one individual level of Parkhurst et al. (2014) model. Here there is the idea that attitudes underpin all four levels of Parkhurst et al. (2014) model. Mollenkopf et al. (2011) and Siren et al. (2015) both place special emphasis on the affective and attitudinal nature of mobility, suggesting this level underpins all other levels of mobility and as such are not necessarily separate as noted by Zieglar and Schwanen (2011) and Musselwhite and Haddad (2010). The memory and imagination element of mobility clearly maps to aesthetic needs in Musselwhite and Haddad (2010) and imaginative mobility in Parkhurst et al. (2014) models. Finally, the substitution of mobility using electronic means, such as the Internet and other forms of technology reducing the need to be literally mobile, maps well to virtual mobility as noted by Parkhurst et al. (2014). Musselwhite (2018b) has placed virtual mobility among all the categories in Musselwhite and Haddad (2010) model rather than it being something separate. For example, virtual mobility can replace literal mobility to the shops with e-shopping and home deliveries, it can help psychosocial needs through social media, and it can meet aesthetic needs through webcams and virtual reality (see Musselwhite 2018a for more discussion).

Point 2: That There Are Legible and Attractive Local Neighbourhoods for Walking (and Cycling) in Later Life

Physical activity is important for health and wellbeing in later life. One of the most important ways to encourage staying active in later life is to build it into everyday life. Walking and cycling ("active travel") as part of daily routines can promote moderate physical activity, improving overall health and wellbeing (Saelens et al. 2003). Yet walking and cycling across the life course is in decline in most wealthy countries and indeed many low-to-middle income countries. In addition to motor vehicles being more accessible and automobile-friendly policies and practices promoted, there are many microsystem and mesosystem barriers to walking and cycling, especially in terms of infrastructure. Matching the infrastructure to the needs of older people is often viewed from the perspective of person-activity-environment fit (Lawton and Nahemow 1973).

The changing physiology of the older person means affordances of infrastructure have to change to enable older people to get out and about. Developing such infrastructure is based on polices, practice and cultural norms, and expectations and the exosystem has to be understood when making changes that might encourage older people to walk and cycle more.

Physical Barriers to Walking and Cycling

Different interactions between physical environmental factors are described well in Alfonzo's Hierarchy of Walking Needs (2005). Alfonzo (2005) describes four levels of need: (1) accessibility (e.g., direct access to local shops and services); (2) safety from crime (e.g., surveillance, hidden spaces); (3) comfort (e.g., smooth pedestrian surfaces, segregation of walking from traffic, amount of benches); and (4) pleasurability (e.g., vegetation, beauty, historic elements). The most important needs have to be satisfied before higher needs so accessibility must be satisfied before safety from crime, comfort, and pleasurability. Each of these can be examined in relation to research on barriers to older people's walking.

A high prevalence and diversity of shops, both grocery and non-grocery, are important determinants as to whether older people walk or not (Cerin et al. 2013). Having services, especially hospital and doctor surgeries, within walking distance from home has been linked to walking among older people (Strath et al. 2012). Recreational accessibility has also been linked to greater walking among older people, including access to places of religious activity (Cerin et al. 2013), parks, and open spaces (Hanibuchi et al. 2011). A good measure of accessibility is walkability. Walkability is a composite environmental measure combining measures such as residential density, land-use mix, and street connectivity (Frank et al. 2010). Residential density is the number of people living in a given area. Land-use mix is the amount of different types of building in an area, for example, residential, commercial, services, offices, and industry. Street connectivity is how well the street is connected to allow direct walking to occur. There is much research that shows high walkability scores, that is, areas with high residential density, high land-use mix and high levels of connectivity, correlate with high levels of older adults walking (e.g., Nathan et al. 2014; Van Holle et al. 2014).

Crime and more importantly fear of crime reduces walking and associated physical and social activity. Fear of crime is prompted by a number of neighbourhood disorders in the areas including litter, dog mess, and graffiti (Burholt et al. 2016). The lack of defensible space (the ability for people to be key agents in their own security of their neighbourhood; Newman 1972), with low levels of visibility from neighbouring windows and too many spaces to hide, can also heighten fear of crime and reduce walking in a neighbourhood (see Cozens et al. 2004). In addition, a Dutch study found fewer older people walking on streets that have "blind walls" with no windows, doors, or openings (Borst et al. 2009).

Poorly maintained pavements that become uneven hamper walking (Newton and Ormerod 2007). Poor weather can also be detrimental to walking for older people. Poor surfaces, caused by fallen leaves, rain, ice, or snow, for example, especially if not cleared, are particularly hazardous (Wennberg 2009). There must be good provision of seating and public conveniences (Musselwhite 2014). Perceiving benches would be present in the public realm is correlated with higher walking, especially among the older participants aged over 75 (Barnett et al. 2016; Cerin et al. 2014) and a lack of benches is a significant barrier to walking (Stahl et al. 2008). Poor crossing facilities are also a crucial factor reducing older people's ability to get out and about, resulting in people making large detours to avoid dangerous crossing of roads or being deterred from using the area altogether (Lord et al. 2010; Zijlstra et al. 2007). Not having long enough time to complete a crossing is a major issue for older people. In many countries formalised crossing facilities are timed to allow people to cross the road at around 1.22 m/second before turning the priority back to motor traffic. Musselwhite (2015), using three case study areas in the UK, found 88% of people aged over 65 did not walk at this speed. This increased to 94% of older females over the age of 65. Previous research has found similar results, suggesting older people's average speeds are between 0.7 and 0.9 metres per second (e.g., Asher et al. 2012; Newton and Ormerod 2007). Poor lighting not only increases safety fears among older people but also reduces ability to see where people are walking and is a significant barrier to walking among older people who can have increased difficulties with luminance (Shumway-Cook et al. 2003). High levels of noise from passing traffic or other pedestrians can also be a barrier (Balfour and Kaplan 2002; Burholt et al. 2016). Poor air quality and pollution is also a barrier for older people getting out and about. Smog for example is an issue in many developing countries, but general air quality can be poor in cities across the world (Deguen and Zmirou-Navier 2010). Finally, natural climate can be a barrier. Very high or very cold temperatures and extreme weather including ice or snow, especially if not cleared from pavements or sidewalks, are barriers for older people walking (Burholt et al. 2016; Williams et al. 2012).

There is less research about the relationship between a lack of pleasurable elements of the environment and a lack of walking among older people. There is much written about the preference of older people for spaces with vegetation and parks (Sugiyama and Ward-Thompson 2008) and neighbourhoods with well-maintained gardens and homes, free from litter and graffiti (Burholt et al. 2016).

Research on shared space, where traditional markers demarcating barriers between vehicles and pedestrians are not present, for example a street having no or a reduced kerb, has suggested that older people tend to stay within the location that used to be the old pavement and do not use the carriageway to walk in when shared space is found (Hammond and Musselwhite 2013; Melia and Moody 2013; Musselwhite 2015). People who do use the carriageway to walk in shared space areas are largely young and almost exclusively male (Kaparias et al. 2010; Moody 2011), suggesting the benefits of shared space may be more apparent to certain groups of pedestrians (Melia and Moody 2013). On the whole, older people do not have additional issues using shared space than any other age group did (Hammond and Musselwhite 2013). But, spaces that are too open and wide can also be viewed negatively, even if they are totally pedestrianised, leading to difficulty in finding orientation for those with cognitive impairments or visual difficulties or by creating a lack of space for refuge or sitting (Atkin 2010).

There is much less written about cycling in later life than there is to do with walking. Barriers to cycling inhabit three main areas: personal, environmental, and psychological. In terms of personal barriers, older people often feel they do not possess enough strength or fitness to cycle. This can be compounded by poor infrastructure, for example, Jones et al. (2016) suggest that streets that make cyclists dismount and remount are especially difficult for older people who may perceive they have lost strength and fitness. Poor environments for cycling, similar to poorly designed or maintained pavements for pedestrians, can also be a barrier. Issues of sharing space with vehicles and with other cyclists as well as walkers can also be a barrier, as can ambiguity of space (Jones et al. 2016). Psychological barriers, including perceived norms, impact on propensity to cycle in later life.

Psychological Barriers to Walking and Cycling: Norms and Culture

With a few exceptions, for example the Netherlands, it has become culturally unusual to walk and cycle in many Western countries, especially everyday or utilitarian walking and cycling. Walking and cycling is often seen as a leisure pursuit, to do in terms of a discretionary activity (Musselwhite 2017). Fewer journeys are made on foot or by bicycle to the shops, to access services, to visit post offices, hospitals, doctors, and similar. People are more likely to walk or cycle if they believe other people like themselves do that activity. In many countries, the stereotype of an older person is not one who does much cycling. In terms of walking, recreational walking is viewed as highly desirable and likely among older people, but walking for a purpose, to go shopping, to go to the doctors, and so on, is viewed as unusual among that age group. Having a dog that needs walking is highly correlated with walking for older people, giving people a sense of legitimacy to walk to the park (Musselwhite 2017). Older people are more likely to walk if they get social support from friends or family (Booth et al. 1997; Booth et al. 2000; Giles-Corti and Donavan 2002). The difference in cultures and cycling rates among older people are also based on cultural norms; how normal is it to see older people on bicycles, for example differs between Denmark, where it is normal and expected, and the UK, where it is an exception (Pucher and Beuler 2008, 2012).

Design of public spaces can exclude groups of people. Older people can feel particularly unable to use public space that lack benches and public conveniences. They also do not always wish to share space that is loud and noisy or inhabited by lot of people, especially younger people who they may feel threatened by. Sometimes, this exclusion occurs simply as the designer has not thought about older people's needs and issues. In such cases there is a need to involve older people in the design of the public realm. On other occasions, the exclusion is deliberate; increasing privatisation of space can design elements of the public realm to create more commercial interactions, and commercial interests of the landowner and tenants are placed above those of the individual. An acute example is found in many Western cities or towns where deliberate lack of places to sit or lack of public conveniences in the public realm force people into cafes or coffee shops, forcing people into commercial transactions just in order to sit down or use the toilet. Ageist stereotypes may also work to keep older people out of certain public realm spaces that the landowner wants to keep looking young, vibrant, and fresh. Across many Western and developing cities, redevelopment and redesign of city centres, for example, are often geared around economic growth with the stereotypical view of a vibrant young wealthy workforce. Hence, homes and commercial space are at best developed for a mythical average person, a hypermobile worker with no dependents and at worst developed for the younger affluent worker, excluding the older person from living in that space.

Towards Convivial Spaces for Walking and Cycling in Later Life

Building on the work by Alves et al. (2008), Sugiyama and Ward-Thompson (2007, 2008), and Musselwhite (2014, 2016, 2018a) and adopting the principles of good design, it is possible to design streets around recognised objectives of urban design as set out by Centre for Architecture and the Built Environment (CABE 2011) (see Table 9.2). Space needs to be safe and accessible, so people feel safe to use it while interacting with other people and with vehicles; legible, so people feel

Table 9.2 Designing streets for older people based on CABE (2011) principles (Musselwhite 2014, 2016, 2018a)

(1) Safe and accessible space—feel you are safe there Sharing space—feel you have room to move, space to dwell for safe interactions with other users	**Ease of movement**	Movement should be enhanced for all users, along with permission to stop and dwell through benches and places to lean and creating focal points to commune at including fountains, works of art, sculptures, memorials or trees, gardens, and other greenery
(2) Legible place Psychological attachment and legitimacy—feel you should be there	**Legibility**	Area should be designed in a way that is easy to understand and interpret, not just with signage but with other visual and tactile cues as well to help determine legitimacy in activity and determine use
	Adaptability	The place should be built to adapt to changes in the needs of users, policy, and legislation over time
	Diversity and choice	Allowing area to be used by a large variety of individuals and uses, with minimum exclusion
(3) Distinctive and aesthetically pleasing—somewhere you want to go and spend time—feel you want to be there	**Character**	Streets should have character and reflect local identify, history, and culture. Utilising local art and architecture can help enhance distinct and unique character and identity
	Continuity and enclosure	Where public and private spaces are easily distinguished
	Quality public realm	Good quality materials easily maintained and replaced

that they can legitimately use them. They should be distinctive and aesthetically pleasing, so that there is some desire to want to use them.

Point 3: That There Is Accessible and Attractive Public and Community Transport to Connect Neighbourhoods and Communities

Public buses play an important part in connectivity for older people, especially those who have given up driving. Bus use is especially high among older people where there is concessionary or free fares, as in the UK. Not only does the bus keep people connected, bus use is also correlated with health and wellbeing, being a protective factor in obesity for older people (Webb et al. 2011). A report from Greener Journeys (2014) suggested that every £1 spent on subsidising travel for older people, returned about £2.87 to the economy in terms of increased access to shops, services, activities, and enabling support for others.

Barriers to Using the Bus

There are still many barriers to using a bus, even if it is free, that prevent or make it difficult for older people to use it. Gilhooly et al. (2002) found that the highest barrier to public transport use amongst older people was personal security in the evening and at night (79.8% of people over 70s agreed), followed by transport running late and having to wait (see Table 9.3). A report using accompanied journeys in London highlighted similar problems for older people including crowds at the bus stop or on the bus, prams taking up the seats or area at the front of the bus, steps up to the bus being too high (or driver stopping too far from the kerb), and fear of falling over when the bus moves off (TfL 2009). Broome et al. (2010) in an Australian study found that for older people, driver friendli-

Table 9.3 Ten most frequent barriers for respondents aged over 70 years, with the proportion of that age group who reported each as a "problem"

Problems	% aged over 70 agreeing
Personal security on the bus in evening and night	79.8%
Public transport running late	68.3%
Having to wait	68%
Carrying heavy loads	66.3%
Potential cancellations	66%
Other passengers behaviour	63.5%
Lack of cleanliness	53.8%
Being out in bad weather	53.8%
Having to change transport	53.3%
Can't travel where I want to	50%
Can't travel when I want to	48.1%

After Gilhooly et al. (2002)

ness, ease of entry/exit, and information usability were prioritised barriers and facilitators for older people. Age UK London (2011) quantified this by surveying bus driving behaviour in 550 journeys in inner London and 541 journeys in outer London. In 42% of cases, passengers were not given enough time to sit down before the bus was driven away from the stop. In 25% of the cases the bus did not pull up tight to the kerb at the bus stop.

Public Transport Norms

One of the major barriers associated with using public transport, such as buses, is the anxiety over the norms of use. It may be, for example, that the user has not used a bus in years. Formal information provision on buses has improved immensely over the past few years, more information than ever before is presented on timetables including real-time and en-route bus stop information. However, older people are anxious about the norms, for example, the normal departure time (is it sooner than is advertised), what times of day are less busy, is there seat availability, are buses accessible, how much can be carried?

One element that stands out frequently is staff attitudes, for example, the bus driver can make or break an older person's journey. A sympathetic driver attuned to older people's needs, who waits for the passenger to sit down before driving off is invaluable. So too is a "cheery" driver who passes the time of day with the older person. Many bus companies have begun to train bus drivers to be sympathetic to the needs of older people in this respect. Older people also want a driver to be friendly, knowledgeable, and helpful to provide information if needed and to be chatty; older people like to get to know drivers they see regularly. It is similar on trains where station and train staff attitude is crucial to successful journeys and the support needs to include practical help with luggage, direction, and train times but also extend to staff having a positive attitude to performing such duties.

Improving Bus Services

Broome et al. (2013) researched how to improve services and overcome barriers for older people using buses in Queensland, Australia, using Brisbane as a control. Seven priorities were implemented, which resulted in increased satisfaction and maintained use among older people:

1. Accessibility, for example, low-floor buses to make boarding and alighting easier;
2. Age-friendly training for bus drivers (see also Broome 2010);
3. Frequent buses and a call for evening and weekend services, in particular;
4. Bus stops close to homes and destinations. Broome et al. (2013) suggest a system should aim to stop within 200 m of residences and destinations. Bus stops need to be of good quality too; they should provide shelter and aid boarding and alighting;
5. Accessible pedestrian infrastructure is important; footpaths and pedestrian crossing are part of the journey too;
6. Providing training and information for older people about how to use buses. To help older people use public transport and get used to the norms, travel training or buddying are sometimes provided. Reflective group work would be beneficial where older people discuss giving up driving perhaps alongside others who have recently given up driving. The group could provide both emotional and practical support. Practical support could include the ability to share lifts in taxis and travel together on buses and to get together for discretionary travel for days out as a group (Musselwhite 2010). Membership could be continuous rather than a programme or cycle of support as is found in an Australian support group (see Liddle et al. 2004, 2006, 2008). Travel buddying could accompany such a group where people new to a mode of transport are accompanied by an expert user. This can be popular for some and again can help overcome the anxiety of travelling alone and gain valuable understanding of the social norms;
7. Bus systems need to provide access to destinations of interest to older people themselves, and Broome et al. (2013) propose that older people should be more involved in design of routes.

Community Transport

As an alternative to conventional public bus services, there can be provision of specialist transport services, often operating door to door for people who cannot access public or private transport, known as specialist transport service or community transport or transit. Such services often run on demand and not usually to a scheduled timetable and are usually run by a third sector or charitable organisation. Such services can provide a lifeline for older people who would otherwise not be able to get out and about. They keep older people connected to services, shops, groups, and clubs and offer an opportunity to socialise with other passengers (ECT (Ealing Community Transport) 2016). There are direct improvements on people's health through affording greater access to General Practitioner (GP) and hospital services and fewer missed appointments, improved diagnosis, and therefore lower healthcare costs (ECT 2016). It can also mean people are discharged earlier as they have access between hospital appointments and home (ECT Charity 2016). Importantly, drivers can act as informal carers and can help identify early warning signs of illness or of loneliness and isolation, as well as offering social support to the passengers (ECT Charity 2016).

There are some barriers to community transport use (Musselwhite 2018c). Services are very dependent on third sector and charity provision and as such provision can be somewhat fragmented across the country. Often the service can be based around a few key individuals, or in some cases a sole individual, running the service, often on very little money or in some cases as volunteers, and if they leave or the funding stop, the service may cease. The service is often provided around the demands of a number of passengers and, as such, cannot compete with the on-demand system of a private car or taxi. Since money is often an issue, the type of transport can often be old and of poor quality. People who may well benefit from such a service can sometimes feel the service is not for people like them; there is sometimes the perception that it is for people with disabilities rather than for everyone with accessibility issues. Frequently, there is a lack of information and as a result much misunderstanding of the service (Parkhurst et al. 2014; Ward et al. 2013). Journeys typically are based around providing transport to shops, services, and doctors and hospitals, but there needs to be more "discretionary" journeys provided to places of leisure and fun (Musselwhite 2017).

Point 4: That There Are Safe, Supportive, and Inclusive Age-Friendly Transport Strategy, Policy, and Plans
Transport policy across many countries is driven by economic interests. As a result, they often centre on commuting and travel for work, which results in transport policy centred on inter- and intra-urban, 9–5 mobility. Older people are less likely to fit this pattern of work and hence can be negatively impacted by such policy. Policy very rarely includes details on individual differences, such as age. Despite this there is much research to say that mobility and transport use can change and alter dramatically at key transition points (Avineri and Goodwin 2010), including those more likely to be faced in later life, such as retirement from full-time employment and the onset of acute or chronic conditions. To make transport policy more relevant to older people, it needs to include a life-course perspective.

Transport policy has a chance to help make mobility easier for older people. In the UK, concessionary or free bus travel does this. Integrated transport policy has been another attempt to improve door-to-door accessibility using public transport, reducing issues of interchange, bringing together distance, timetables, ticketing, information, and service design between modes (see Table 9.4).

Conclusion

This four-point plan can aid older people's mobility beyond the car. It shows clearly different barriers and enablers to mobility for older people focusing between macro and microscopic layers and showing different actors' and agents' roles in making this happen. At the individual level, there is a need to address needs and motivations for mobility, examining the importance of practical and also the affective elements of mobility. As people age, they spend more time in their local neighbourhood and hence

Table 9.4 Integrated transport and examples

Type of integration	Description	Example
Locational integration	Being able to easily change between transport modes (using Interchanges)—this is about services connecting in space	Over the last 20 years, the railways have developed "Parkway" stations, for example, Luton Parkway, Bristol Parkway, Southampton Airport Parkway, and East Midlands Parkway. A number of these also double as bus/coach/air and rail interchanges as well
Timetabling integration	Services at an interchange connect in time	In San Francisco's Bay Area, the Bay Area Rapid Transit (BART) Metro links into local bus services at suburban interchange stations. Buses are scheduled to depart five minutes after the BART train arrives. However, a key aspect here is that both BART and the buses are state owned and timetable integration is a result of a policy decision
Ticketing integration	Not needing to purchase a new ticket for each leg of a journey	Integrated ticketing already occurs in some large cities, for example, London Oyster Card and Plus Bus
Information integration	Not needing to enquire at different places for each stage of a trip—or that different independent sources are easily connected	Main line rail stations are beginning to provide poster displays of bus services from the station, the location of bus stops, and a street map of the area within about a five-minute walk. Real-time transport information is starting to be provided
Service design integration	That the legal, administrative, and governance structures permit/encouraging integration	This can easily happen when responsibility and planning for transport are devolved to local areas, for example, Transport for London

After Potter (2010)

this can be a major barrier or enabler to being able to leave home. Almost all journeys begin and end in the neighbourhood and without a vehicle, so it is vital to get this right. Movement to further away places connecting the immediate neighbourhood to other communities needs to be age friendly. Supporting public and community transport must provide this role beyond the car. The neo-liberal approach of leaving transport provision to the market can lead to fragmented services, where provision is placed mainly around the need to work and to grow the economy. Hence there is a dominance of provision for private mobility, especially for cars and goods, leaving those on the margins of the mainstream, for example, older people, at risk of exclusion and isolation. Therefore, all these levels need top-down support from policy to help integrate the different actors and agents involved from the different transport providers and smooth over any social exclusion occurring for older people as a result of disjointed transport provision.

Traditionally, research has focused on practical barriers to mobility in later life. This centres on the environment-competence mismatch, where improvements in the infrastructure environment will reduce deficit and improve mobility for older people. Many of the barriers to mobility in older age are psychological in nature. Mobility is related more than simply to practical issues and outcomes and involves an affective, emotive, and psychosocial component. Mobility helps foster a sense of independence and normality involving a form of impression management and is related to self-identity and self-esteem. Since mobility sits within wider social contexts, social norms play an important role in defining accessibility to transport in later life. It has become so normal to use the car in many Western cultures that using other modes of transport are seen as unusual. Having to give up driving means older people have to engage with these alternative forms of transport, marking them as different or unusual; a barrier to integrating such practices into daily routines. It is sadly more unusual to be seen to walk or cycle than ever before in many Western countries and even more so if you are an older person. The norms of using a bus are often not known amongst a cohort who have spent most of their adult life driving, meaning using buses for the first time can be challenging. Planning for mobility for older people post-driving needs to take into account these psychological issues. To fully embrace transport in its social context, a change of research question and focus is required. It also requires a change in attitude from policymakers and practitioners to embrace such change. It may require practitioners and politicians to be more involved in the research process and to actively engage in the research findings in order to generate recommendations that will have an impact. In order to understand transport from the human perspective and answer the transport issues of the day, there is a need for a more human-centred approach to transport studies.

References

Adler, G., & Rottunda, S. (2006). Older Adults' Perspectives on Driving Cessation. *Journal of Aging Studies, 20*(3), 227–235.

Alfonzo, M. A. (2005). To Walk or Not to Walk? The Hierarchy of Walking Needs. *Environment & Behavior, 37*(6), 808–836.

Alves, S., Aspinall, P., Ward Thompson, C., Sugiyama, T., Brice, R., & Vickers, A. (2008). Preferences of Older People for Environmental Attributes of Local Parks: The Use of Choice-Based Conjoint Analysis. *Facilities, 26*(11/12), 433–453.

Asher, L., Aresu, M., Falaschetti, E. A., & Mindell, J. (2012). Most Older Pedestrians Are Unable to Cross the Road in Time: A Cross-Sectional Study. *Age and Ageing, 41*, 690–694.

Atkin, R. (2010). *Sight Line. Designing Better Streets for People with Low Vision.* London: Helen Hamlyn Centre, Royal College of Art.

Avineri, E., & Goodwin, P. (2010). *Individual Behaviour Change: Evidence in Transport and Public Health.* London: The Department for Transport.

Balfour, J. L., & Kaplan, G. A. (2002). Neighborhood Environment and Loss of Physical Function in Older Adults: Evidence from the Alameda County Study. *American Journal of Epidemiology, 155*(6), 507–515.

Barnett, A., Cerin, E., Zhang, C. J. P., Sit, C. H. P., Johnston, J. M., Cheung, M. M. C., & Lee, R. S. Y. (2016). Associations Between the Neighbourhood Environment Characteristics and Physical Activity in Older Adults with Specific Types of Chronic Conditions: The ALECS Cross-Sectional Study. *International Journal of Behavioral Nutrition & Physical Activity, 13*. doi:https://doi.org/10.1186/s12966-016-0377-7

Berk, L. E. (2000). *Child Development.* Boston: Allyn and Bacon.

Booth, M. L., Bauman, A., Owen, N., & Gore, C. J. (1997). Physical Activity Preferences, Preferred Sources of Assistance, and Perceived Parries to Increased Activity Among Physically Inactive Australians. *Preventative Medicine, 26*, 131–137.

Booth, M. L., Owen, N., Bauman, A., Clavisi, O., & Leslie, E. (2000). Social-Cognitive and Perceived Environmental Influences Associated with Physical Activity in Older Australians. *Preventive Medicine, 31*, 15–22.

Borst, H. C., De Vries, S. I., Graham, J. M. A., Van Dongen, J. E. F., Bakker, I., & Miedema, H. M. E. (2009). Influence of Environmental Street Characteristics on Walking Route Choice of Elderly People. *Journal of Environmental Psychology, 29*(4), 477–484. https://doi.org/10.1016/j.jenvp.2009.08.002

Bronfenbrenner, U. (1979). *Ecology of Human Development.* Cambridge, MA: Harvard University Press.

Bronfenbrenner, U. (1989). Ecological Systems Theory. *Annals of Child Development, 6*, 185–246.

Bronfenbrenner, U. (2005). *Making Human Beings Human: Bioecological Perspectives on Human Development.* Thousand Oaks: Sage.

Broome, K., Nalder, E., Worrall, L., & Boldy, D. (2010). Age-Friendly Buses? A Comparison of Reported Barriers and Facilitators to Bus Use for Younger and Older Adults. *Australias Journal of Ageing, 29*(1), 33–38.

Broome, K., Worrall, L., Fleming, J., & Boldy, D. (2013). Evaluation of Age-Friendly Guidelines for Public Buses. *Transportation Research Part A: Policy and Practice, 53*, 68–80.

Burholt, V., Roberts, M. S., & Musselwhite, C. B. A. (2016). Older People's External Residential Assessment Tool (OPERAT): A Complementary Participatory and Metric Approach to the Development of an Observational Environmental Measure. *BMC Public Health, 16*, 1022.

CABE (Centre for Architecture and the Built Environment). (2011). *Seven Principles of Good Design.* http://webarchive.nationalarchives.gov.uk/20110118095356/http:/www.cabe.org.uk/councillors/principles. Last Accessed 11 Jan 2018.

Cerin, E., Lee, K. Y., Barnett, A., Sit, C. H. P., Cheung, M. C., & Chan, W. M. (2013). Objectively-Measured Neighborhood Environments and Leisure-Time Physical Activity in Chinese Urban Elders. *Preventive Medicine, 56*(1), 86–89.

Cerin, E., Sit, C. H. P., Barnett, A., Johnston, J. M., Cheung, M. C., & Chan, W. M. (2014). Ageing in an Ultra-Dense Metropolis: Perceived Neighbourhood Characteristics and Utilitarian Walking in Hong Kong Elders. *Public Health Nutrition, 17*(1), 225–232.

Cozens, P. M., Neale, R., Whitaker, J., & Hillier, D. (2004). Tackling Crime and Fear of Crime While Waiting at Britain's Railway Stations. *Journal of Public Transportation, 7*(3), 23–41.

Davey, J. A. (2007). Older People and Transport: Coping Without a Car. *Ageing and Society, 27*, 49–65.

Deguen, S., & Zmirou-Navier, D. (2010). Social Inequalities Resulting from Health Risks Related to Ambient Air Quality—A European Review. *European Journal of Public Health, 20*, 27–35.
ECT (Ealing Community Transport). (2016). *Why Community Transport Matters? ECT Transport Report*. Greenford: ECT. Available at https://ectcharity.co.uk/files_uploads/ECT_Why_community_transport_matters_Final_version4.pdf. Last accessed 27 July 2018.
Edwards, J. D., Perkins, M., Ross, L. A., & Reynolds, S. L. (2009). Driving Status and Three Year Mortality Among Community-Dwelling Older adults. *Journal of Gerontology Series A: Biological Sciences and Medical Sciences, 64*, 300–305.
Fonda, S. J., Wallace, R. B., & Herzog, A. R. (2001). Changes in Driving Patterns and Worsening Depressive Symptoms Among Older Adults. *The Journal of Gerontology, Series B: Psychological Sciences and Social Sciences, 56*(6), 343–351.
Frank, L. D., Sallis, J. F., Saelens, B. E., Leary, L., Cain, K. L., Conway, T. L., & Hess, P. M. (2010). The Development of a Walkability Index: Application to the Neighborhood Quality of Life Study. *British Journal of Sports Medicine, 44*(13), 924–933. https://doi.org/10.1136/bjsm.2009.058701
Giles-Corti, B., & Donovan, R. J. (2002). The Relative Influence of Individual, Social and Physical Environmental Determinants of Physical Activity. *Social Science Medicine, 54*(12), 1793–1812.
Gilhooly, M., Hamilton, K., & O'Neill, M. (2002). *Transport and Ageing Extending Quality of Life for Older People Via Public and Private Transport*. Glasgow: Glasgow Caledonian University.
Greener Journeys. (2014). *Concessionary Travel Costs and Benefits September 2014*. Available at: www.greenerjourneys.com/bus-pass/research/. Accessed 11 Jan 2018.
Haglund, M., & Åberg, L. (2000). Speed Choice in Relation to Speed Limit and Influences from Other Drivers. *Transportation Research Part F, 3*, 39–51.
Hammond, V., & Musselwhite, C. B. A. (2013). The Attitudes, Perceptions and Concerns of Pedestrians and Vulnerable Road Users to Shared Space: A Case Study from the UK. *Journal Of Urban Design, 18*(1), 78–97.
Hanibuchi, T., Kawachi, I., Nakaya, T., Hirai, H., & Kondo, K. (2011). Neighborhood Built Environment and Physical Activity of Japanese Older Adults: Results from the Aichi Gerontological Evaluation Study (AGES). *BMC Public Health, 11*, 657.
Holt, S., Buckley, H., & Whelan, S. (2008). The Impact of Exposure to Domestic Violence on Children and Young People: A Review of the Literature. *Child Abuse & Neglect, 32*, 797–810.

Jones, T., Chatterjee, K., Spinney, J., Street, E., Van Reekum, C., Spencer, B., Jones, H., Leyland, L. A., Mann, C., Williams, S., & Beale, N. (2016). *Cycle BOOM. Design for Lifelong Health and Wellbeing. Summary of Key Findings and Recommendations*. Oxford: Oxford Brookes University.

Kaparias, I., Bell, M. G. H., Miri, A., Cheng, S., Greensted, J., & Taylor, C. (2010). *Modelling the Willingness of Pedestrians to Share Space with Vehicles* (pp. 1–12). Paper Presented at the UTSG Conference, Imperial College, London.

Keating, N., & Phillips, J. E. (2008). A Critical Human Ecology Perspective on Rural Ageing. In N. Keating (Ed.), *Rural Ageing: A Good Place to Grow Old?* (pp. 1–10). Bristol: Policy Press.

Kulik, L., & Rayyan, F. (2006). Relationships Between Dual-Earner Spouses, Strategies for Coping with Home-Work Demands and Emotional Well-Being: Jewish and Arab-Muslim Women in Israel. *Community, Work & Family, 9*, 457–477.

Lawton, M. P. (1999). Environmental Taxonomy: Generalizations from Research with Older Adults. In S. Friedman & T. Wachs (Eds.), *Measuring Environment Across the Lifespan* (pp. 91–124). Washington, DC: American Psychological Association.

Lawton, M. P., & Nahemow, L. (1973). Ecology and the Aging Process. In C. Eisoderofer & M. P. Lawton (Eds.), *Psychology of Adult Development and Aging* (pp. 619–674). Washington, DC: American Psychological Association.

Liddle, J., Carlson, G., & McKenna, K. (2004). Using a Matrix in Life Transition Research. *Qualitative Health Research, 14*, 1396–1417.

Liddle, J., McKenna, K., & Bartlett, H. (2006). Improving Outcomes for Older Retired Drivers: The UQDRIVE Program. *Australian Occupational Therapy Journal, 53*, 1–4.

Liddle, J., Turpin, M., Carlson, G., & McKenna, K. (2008). The Needs and Experiences Related to Driving Cessation for Older People. *British Journal of Occupational Therapy, 71*(9), 379–388.

Ling, D. J., & Mannion, R. (1995). Enhanced Mobility and Quality of Life of Older People: Assessment of Economic and Social Benefits of Dial-a-Ride Services. In *Proceedings of the Seventh International Conference on Transport and Mobility for Older and Disabled People* (Vol. 1). London: DETR.

Lord, S. E., Weatherall, M., & Rochester, L. (2010). Community Ambulation in Older Adults: Which Internal Characteristics Are Important? *Archives of Physical Medicine and Rehabilitation, 91*(3), 378–383.

Marottoli, R. A., Mendes de Leon, C. F., Glass, T. A., Williams, C. S., Cooney, L. M., Jr., Berkman, L. F., & Tinetti, M. E. (1997). Driving Cessation and Increased Depressive Symptoms: Prospective Evidence from the New Haven EPESE. *Journal of the American Geriatric Society, 45*(2), 202–206.

Marottoli, R. A., Mendes de Leon, C. F., Glass, T. A., Williams, C. S., Cooney, L. M., & Berkman, L. F. (2000). Consequences of Driving Cessation: Decreased Out-of-Home Activity Levels. *Journals of Gerontology: Series B, Psychological Sciences and Social Sciences, 55B*(6), 334–340.

Metz, D. (2000). Mobility of Older People and Their Quality of Life. *Transport Policy, 7*, 149–152.

Mezuk, B., & Rebok, G. W. (2008). Social Integration and Social Support Among Older Adults Following Driving Cessation. *Journal of Gerontology Social Science, 63B*, 298–303.

Moody, S. (2011). *How Do Pedestrians Move in a Shared Space Scheme with High Traffic Flows? Is the Pedestrian Empowered or Does a Perception of Risk Give Rise to Anxiety and a Disincentive to Share the Street with Vehicles?* MSc Transport Planning Dissertation. University West of England, Bristol, UK.

Moody, S., & Melia, S. (2013). Shared Space – Research, Policy and Problems. *Proceedings of the Institution of Chartered Civil Engineers – Transport*. Available at: www.icevirtuallibrary.com/content/article/10.1680/tran.12.00047. Accessed 27 July 2018.

Mollenkopf, H., Hieber, A., & Wahl, H.-W. (2011). Continuity and Change in Older Adults' Perceptions of Out-of-Home Mobility Over Ten Years: A Qualitative-Quantitative Approach. *Ageing & Society, 31*(5), 758–781.

Musselwhite, C. B. A. (2010). The Role of Education and Training in Helping Older People to Travel After the Cessation of Driving. *International Journal of Education and Ageing, 1*(2), 197–212.

Musselwhite, C. (2014). Designing Public Space for Older People. *Generations Review, 24*(3), 25–27.

Musselwhite, C. B. A. (2015). Environment-Person Interactions Enabling Walking in Later Life. *Transport Planning & Technology, 38*(1), 44–61.

Musselwhite, C. B. A. (2016). Vision for an Age Friendly Transport System in Wales. *EnvisAGE, Age Cymru, 11*, 14–23.

Musselwhite, C. B. A. (2017). Exploring the Importance of Discretionary Mobility in Later Life. *Working with Older People, 21*(1), 49–58.

Musselwhite, C. B. A. (2018a). Creating a Convivial Public Realm for an Ageing Population. Being a Pedestrian and the Built Environment. In C Musselwhite (Ed.), *Transport, Travel and Later Life* (Transport and Sustainability, Vol. 10). Bingley: Emerald Publishing Limited, pp.129 - 137

Musselwhite, C. B. A. (2018b). Virtual and Imaginative Mobility: How Do We Bring the Outside Indoors and What Happens When Mobility Is No Longer Available? In C. Musselwhite (Ed.), *Transport, Travel and Later Life (Transport and Sustainability, Volume 10)* (pp. 197–205). Bingley: Emerald Publishing Limited.

Musselwhite, C. B. A. (2018c). Public and Community Transport. In C. Musselwhite (Ed.), *Transport, Travel and Later Life* (Transport and Sustainability, Vol. 10, pp. 117–128). Bradford: Emerald Publishing Limited.

Musselwhite, C. B. A., & Haddad, H. (2010). Mobility, Accessibility and Quality of Later Life. *Quality in Ageing and Older Adults, 11*(1), 25–37.

Musselwhite, C. B. A., & Shergold, I. (2013). Examining the Process of Driving Cessation in Later Life. *European Journal of Ageing, 10*(2), 89–100.

Musselwhite, C. B. A., Avineri, E., & Susilo, Y. O. (2014). Legitimising Risk Taking: Articulating Dangerous Behaviour on the Road. *Transportation Planning and Technology, 37*(1), 62–82.

Nathan, A., Wood, L., & Giles-Corti, B. (2014). Perceptions of the Built Environment and Associations with Walking Among Retirement Village Residents. *Environment & Behavior, 46*(1), 46–69.

Newman, O. (1972). *Defensible Space*. New York: Macmillan.

Newton, R., & Ormerod, M. (2007). *The Design of Streets with Older People in Mind: Materials of Footways and Footpaths*. IDGO Design Guide. Available at: http://www.idgo.ac.uk/design_guidance/factsheets/materials_footways_footpaths.htm. Last Accessed 11 Jan 2018.

O'Connell, M. (2002). Social Psychological Principles: 'The Group Inside the Person'. In R. Fuller & J. A. Santos (Eds.), *Human Factors for Highway Engineers* (pp. 201–215). Amsterdam: Pergamon.

Ormerod, M., Newton, R., Philips, J., Musselwhite, C., McGee, S., & Russell, R. (2015). *How Can Transport Provision and Associated Built Environment Infrastructure Be Enhanced and Developed to Support the Mobility Needs of Individuals as They Age? Future of an Ageing Population: Evidence Review*. London: Foresight, Government Office for Science.

Parkhurst, G., Galvin, K., Musselwhite, C., Phillips, J., Shergold, I., & Todres, L. (2014). Beyond Transport: Understanding the Role of Mobilities in Connecting Rural Elders in Civic Society. In C. Hennesey, R. Means, & V. Burholt (Eds.), *Countryside Connections: Older People, Community and Place in Rural Britain* (pp. 125–156). Bristol: Policy Press.

Peace, S., Holland, C., & Kellaher, L. (2006). *Environment and Identity in Later Life*. Maidenhead: Open University Press.

Peel, N., Westmoreland, J., & Steinberg, M. (2002). Transport Safety for Older People: A Study of Their Experiences, Perceptions and Management Needs. *Injury Control & Safety Promotion, 9*, 19–24.

Phillipson, C., & Walker, A. (Eds.). (1986). *Ageing and Social Policy: A Critical Assessment*. Aldershot: Gower.

Phillipson, C., Allan, G., & Morgan, D. (2000). *Social Networks and Social Exclusion*. Aldershot: Ashgate Books.

Potter, S. (2010, January 5–7). Transport Integration – An Impossible Dream? In *Proceedings of the Universities Transport Studies Group Annual Conference*. University of Plymouth.

Pucher, J., & Buehler, R. (2008). Making Cycling Irresistible: Lessons from The Netherlands, Denmark and Germany. *Transport Reviews, 28*(4), 495–528.

Pucher, J., & Buehler, R. (2012). *City Cycling*. Cambridge, MA: MIT Press.

Ragland, D. R., Satariano, W. A., & MacLeod, K. E. (2005). Driving Cessation and Increased Depressive Symptoms. *The Journals of Gerontology Series A: Biological Sciences and Medical Sciences, 60*, 399–403.

Saelens, B. E., Sallis, J. F., & Frank, L. D. (2003). Environmental Correlates of Walking and Cycling: Findings from the Transportation, Urban Design, and Planning Literature. *Annals of Behaviour Medecine, 25*, 80–91.

Schlag, B., Schwenkhagen, U., & Trankle, U. (1996). Transportation for the Elderly: Towards a User-Friendly Combination of Private and Public Transport. *IATSS Research, 20*(1), 75–82.

Shergold, I., Parkhurst, G., & Musselwhite, C. (2012). Rural Car Dependence: An Emerging Barrier to Community Activity for Older People? *Transport Planning & Technology, 35*(1), 69–85.

Shumway-Cook, A., Patla, A., Stewart, A., Ferrucci, L., Ciol, M. A., & Guralnik, J. M. (2003). Environmental Components of Mobility Disability in Community-Living Older Persons. *Journal of American Geriatrics Society, 51*(3), 393–398.

Siren, A., & Hakamies-Blomqvist, L. (2009). Mobility and Well-Being in Old Age. *Topics in Geriatric Rehabilitation, 25*, 3–11.

Siren, A., Hjorthol, R., & Levin, L. (2015). Different Types of Out-of-Home Activities and Well-Being Amongst Urban Residing Old Persons with Mobility Impediments. *Journal of Transport and Health, 2*(1), 14–21.

Stahl, A., Carlsson, G., Hovbrandt, P., & Iwarsson, S. (2008). "Let's Go for a Walk!": Identification and Prioritisation of Accessibility and Safety Measures Involving Elderly People in a Residential Area. *European Journal of Ageing, 5*(3), 265–273.

Strath, S. J., Greenwald, M. J., Isaacs, R., Hart, T. L., Lenz, E. K., Dondzila, C. J., & Swartz, A. M. (2012). Measured and Perceived Environmental Characteristics Are Related to Accelerometer Defined Physical Activity in Older Adults. *International Journal of Behavioral Nutrition and Physical Activity, 9*(40), 40.

Sugiyama, T., & Ward Thompson, C. (2007). Outdoor Environments, Activity and the Well-Being of Older People: Conceptualising Environmental Support. *Environment and Planning A, 39*, 1943–1960.

Sugiyama, T., & Ward Thompson, C. (2008). Associations Between Characteristics of Neighbourhood Open Space and Older People's Walking. *Urban Forestry & Urban Greening, 7*(1), 41–51.

TfL. (2009). *Older People's Experience of Travel in London*. London: Transport for London.

Van Holle, V., Van Cauwenberg, J., Van Dyck, D., Deforche, B., Van De Weghe, N., & De Bourdeaudhuij, I. (2014). Relationship Between Neighborhood Walkability and Older Adults' Physical Activity: Results from the Belgian Environmental Physical Activity Study in Seniors (BEPAS Seniors). *International Journal of Behavioral Nutrition & Physical Activity, 11*, 110.

Ward, M., Somerville, P., & Bosworth, G. (2013). 'Now Without My Car I Don't Know What I'd Do': The Transportation Needs of Older People in Rural Lincolnshire. *Local Economy, 28*(6), 553–566.

Webb, E., Netuveli, G., & Millett, C. (2011). Free Bus Passes, Use of Public Transport and Obesity Among Older People in England. *Journal of Epidemiology and Community Health, 66*(2), 176–180.

Webber, S. C., Porter, M. M., & Menec, V. H. (2010). Mobility in Older Adults: A Comprehensive Framework. *Gerontologist, 50*(4), 443–450.

Wennberg, H. (2009). *Walking in Old Age: A Year-Round Perspective on Accessibility in the Outdoor Environment and Effects of Measures Taken*. PhD Thesis, Lund University. Available at: http://www.fot.se/documents/Wennberg_-_Walking_in_old_age_KAPPAN_-_2009.pdf. Last Accessed 11 Jan 2018.

Williams, K., Gupta, R., Smith, I., Joynt, J., Hopkins, D., Bramley, G., Payne, C., Gregg, M., Hambleton, R., Bates-Brkljac, N., Dunse, N., & Musselwhite, C. (2012). *Suburban Neighbourhood Adaptation for a Changing Climate (SNACC). Final Report*. Bristol: University of the West of England.

Windsor, T. D., Anstey, K. J., Butterworth, P., Luszcz, M. A., & Andrews, G. R. (2007). The Role of Perceived Control in Explaining Depressive Symptoms Associated with Driving Cessation in a Longitudinal Study. *The Gerontologist, 47*, 215–223.

Ziegler, F., & Schwanen, T. (2011). I Like to Go Out to Be Energised by Different People: An Exploratory Analysis of Mobility and Wellbeing in Later Life. *Ageing and Society, 31*(5), 758–781.

Zijlstra, G. A., van Haastregt, J. C., van Eijk, J. T., van Rossum, E., Stalenhoef, P. A., & Kempen, G. I. (2007). Prevalence and Correlates of Fear of Falling, and Associated Avoidance of Activity in the General Population of Community Living Older People. *Age and Ageing, 36*(3), 304–309.

10

Resisting Negative Social Representations of Ageing

Katie Wright-Bevans and Michael Murray

Psychological research on ageing has often been concerned with the internal psychic processes involved in growing older and the individual characteristics of older people with less consideration given to their location with the broader social world. This chapter deliberately adopts a societal psychological perspective locating older people in a social, symbolic, and material world that frequently negatively stereotypes old age. However, within such a world it is possible to develop a range of strategies of resistance to such negative stereotyping. Two overlapping theories guide our exploration of these processes: social representation and narrative theory, both theories of meaning-making which interconnect (Murray 2002). Social representation and narrative theory provide a way to combat the individualism of much psychological theory, research, and practice. They do so by providing a way to understand the conflicting meanings of ageing, how certain definitions of ageing come to dominate, the consequences of those meanings for older people, and, finally, how older people

K. Wright-Bevans (✉) • M. Murray
School of Psychology, Keele University, Newcastle-under-Lyme, UK
e-mail: k.wright.bevans@keele.ac.uk; m.murray@keele.ac.uk

E. Peel et al. (eds.), *Psychologies of Ageing*,
https://doi.org/10.1007/978-3-319-97034-9_10

and others can resist negative stereotypes of ageing. This chapter discusses how both theories contribute to an understanding of ageing with reference particularly in the latter half to several research examples.

In western culture, ageing is often associated with decline and disengagement (Calasanti 2008). However, social, medical, and technological advances have necessitated a redefining of what it means to age as the nature of later life changes. Traditional psychological research has been concerned with the identification of various deficits in functioning among older people. The reductionist orientation of this approach carries with it an implicit blaming of those older people who cannot meet the criteria for successful ageing (McCann et al. 2008). Conversely, social representation theory locates older people within a symbolic web of conflicting social meanings that provide opportunities for resistance. Social representations act as a social code, communicating norms, assumptions, and values about ageing. In this chapter, we examine ageing not as an individual experience, separate from culture and context, nor as a societal discourse evading agency and resistance but instead as social and cultural phenomena, embedded with common-sense ideas, which are often rooted in fears, myths, and stereotypes. This common sense is evident in how people age, talk about ageing, and interact with older people. Meanings of ageing are context-specific and evident in practice, culture, and language as well as being shaped or resisted by these communicative acts.

We explore the role of power and powerful social agents (e.g., health professionals) in legitimising and perpetuating hegemonic (i.e., politically powerful) social representations of ageing as well as the capacity of individuals and groups for empowerment, resistance, and positive social change. In the latter section of this chapter, we draw on a range of research examples to illustrate different strategies by which individuals, groups, and communities can challenge dominant negative narratives and social representations of ageing.

Social Representation Theory and Ageing

Social representations shape and define what it means to age and age successfully. They are a communicative tool, which aim to make familiar the unfamiliar (Moscovici 1988); they are localised systems of meaning

(Jovchelovitch 2007). Moscovici (1976: xiii) defined a social representation as:

> a system of values, ideas and practices with a twofold function: first, to establish an order which will enable individuals to orientate themselves in their material and social world and to master it; secondly to enable communication to take place among the members of a community by providing them with a code for social exchange and a code for naming and classifying unambiguously the various aspects of their world and their individual and group history.

In our everyday social interaction, we are involved in the creation, maintenance, and resistance to these social representations.

As a theory of social knowledge and social change, the theory of social representations is able to account for the interplay between social knowledge and social identities (Howarth et al. 2013) and embodiment (O'Connor 2006) and participation (Howarth et al. 2014). Several researchers have studied social representations of health, illness, and ageing. Herzlich's (1974) influential work on social representations of health and illness demonstrated how French people defined health in a holistic way, as engagement in society and defined illness as exclusion and non-participation. Research on older adults' social representations of health and illness (Murray et al. 2003) and of ageing (Moreno et al. 2016) has demonstrated the context-specific nature of these meanings and their importance in shaping the ageing experience. The theory of social representations bridges the gap between medical sociology, social gerontology, and mainstream approaches to the psychology of ageing by offering a social psychology of ageing in which the individual is social and the social is individual.

Crucially, the theory of social representations recognises that knowledge is neither apolitical nor value-free. This characteristic of the theory enables a full account of the role of power in social knowledge, including the roles of the powerful and the powerless. It is the malleable nature of a social representation, which most prominently distinguishes it from the notion of a discourse (Howarth 2006). Discourses similarly recognise the social and political, but whereas discourses of ageing might be described as widespread ideas that bear down on and shape the lives of older people, social representations are subject to ongoing negotiation and resistance through representation. This sensitivity to the negotiated nature of meaning lends to the theory's critical capacity, raising questions of authority, voice, and legitimacy in meaning-making.

Dominant Social Representations of Ageing

Throughout the western world, the dominant social representation of later life is negative. Ageing is represented as a time of dependency, social isolation, and fragility. These negative meanings have maintained currency over time, across generations and remain stable in current commonsense understandings, despite a diversification of experiences of ageing in the twenty-first century. The widespread nature of these negative social representations of ageing has meant, for some older people, a reluctance to identify as an older person (Quéniart and Charpentier 2012), a resistance to being labelled 'old'.

How Meanings Come to Dominate

There is rarely ever only one set of meanings, of social representations, regarding a particular phenomenon. Multiple systems of knowledge circulate and compete, often in tension. The degree of power held by the various social agents involved helps to determine which social representations become reified, hegemonic, and concrete and which are sidelined or suppressed. In diverse and contemporary societies, different groups hold different and competing ideas, norms, and values and these different social representations can and do come into conflict. Within empirical research much attention has turned to lay knowledge and sense-making, particularly with respect to health (Flick et al. 2003), sometimes in order to contrast such lay meaning-making with the knowledge of health professionals (Morant 2006). However, it is equally important to examine meanings of ageing, health, and related concepts among the most powerful. These social representations may be granted greater legitimacy and therefore have greater consequences for older people. The seeming legitimacy of professional and scientific knowledge has warranted its evasion from empirical scrutiny. The idea that scientific and professional knowledge entail social constructions is not a new one (see Kuhn 1970). However, within mainstream psychology and the psychology of ageing more specifically, where interests often lie with the individual experience, less research has critically examined the role of powerful others, their social representations, and the impact of these.

As all knowledge is political (Elcheroth et al. 2011), social representations are political. Their construction and circulation display the power of the social agents that use them. Those groups with greater, more widespread, and more powerful outlets for their voice give greater exposure (and therefore credibility) to their social representations. In the same vein, those with less voice, with fewer opportunities and avenues to be heard, are frequently misunderstood and their experiences, values, and identities become 'othered' (Joffe 1999). In contexts where there is an absence of images of ageing (Twigg 2004), this does not leave a void for positive images to fill but instead an absence of images contributes to dominant social representations of ageing as passive, as 'other', and unwanted in relation to youth as active and desirable. The question of how particular meanings come to dominate is important both in understanding ageing and in identifying ways to facilitate environments in which people can age well. Although social representations are malleable, they do remain relatively stable and must do so in order to carry out their communicative function. These more stable dominant negative social representations are problematic for groups of older people with little instrumental or symbolic power. Where groups are powerless, they often cannot identify means of resisting dominant negative images of ageing.

Social representations that circulate among those with more power, whether they be professionals, policymakers, or popular figures, are not only more stable but also become more easily legitimised and likely to pass as fact. Power, whether instrumental or symbolic, often leads to the reification and legitimisation of a particular system of social representations. Powerful social agents such as governments, professionals, experts, and the mass media monopolise what it means to age and to age well. Dominant negative meanings of ageing are not benign or free from consequence. Instead, such knowledge can function to legitimise and normalise negative or harmful practices, cultures of exclusion, and the unequal distribution of resources. Negative dominant social representations through their circulation become both the cause and consequence of negative attitudes and behaviours towards ageing and older people. Such causes and consequences are evidenced by the social exclusion of older people from particular places and spaces, the invisibility of older people's social, cultural, and economic contribution and citizenship, and also malpractice in care homes. These are all examples of how negative social representations of ageing, perpetuated by the powerful, affect the powerless. These dominant meanings act upon and shape what it means to age. They have tangible consequences for older people. An example to illustrate this is in town and city planning, where older people do not have a voice, are invisible or unheard and resources are limited, facilities to enhance their quality of life, access to places and spaces and cultural engagement are overlooked, misunderstood, or ignored (Buffel et al. 2012). As a result, community spaces, seating, or pathways are often inadequate or inaccessible to many older people. Unfortunately, such actions only serve to perpetuate further those social representations of ageing as dependent and vulnerable.

The Role of Science and Scientific Thinking

Common-sense meaning-making is not confined to lay and professional knowledge but is also evident in scientific thinking (Moscovici and Markova 1998). Flick (1998) stated that when a new phenomenon is the result of a scientific process (rather than a process in a lay community), a different and more concrete level of objectification is observed. The objectification of images of ageing as a period of decline is therefore supported and cemented by an increase in scientific discussion around ageing and cognitive decline. Increased research and scientific discussion (fuelled by attention from the mass media) around cognitive decline (e.g., risk factors associated with dementia) fuel a social representation of ageing as a period of decline. Such medical knowledge, debate, and scientific thinking inevitably contribute to common-sense images of ageing. Language used by those in positions of power, with regard to ageing is not the only contributor to dominant social representations. Social representations not only constitute ideas and values expressed through language but also are evident in action. The practices of experts and professionals feed and shape common-sense understandings around the needs of older people. Harmful or exclusionary practices therefore feed and shape social representations of older people as invisible, passive, or dependent. Research has examined systems of values, ideas, and practices among health professionals (Flick et al. 2003) and social care professionals (Renedo and Jovchelovitch 2007) and demonstrated how professional talk and practice both contribute to the social representations of the populations they work with. Flick et al. (2003) conducted interviews with nurses and general practitioners in which the interview focused on eliciting examples of practice from the participants. Renedo and Jovchelovitch (2007) found that social representations held by front-line workers conflicted with those held by rear-line workers in greater positions of power. Systems of practice and ways of acting may eventually become cultural and social norms within given professions. This normalisation, combined with the status of those involved in such practices, ensures that negative practices are often difficult to combat or resist.

The Role of Institutions

Over time dominant social representations come to show themselves in the very fabric of institutions; they become formalised and act as social facts woven into social institutions, their architecture, geography, cultures, and policy. Social representations come to dictate who can and cannot occupy certain places and spaces, how people should and should not behave, and what people can and cannot say. For many social spaces, social representations contribute to an understanding of whether or not older people should or should not be there. Retirement villages and various residential settings may appeal to some older people as more inclusive spaces in which old age is not othered (see also Chap. 8, this volume). This segregation of older people has been argued as problematic. Hagestad and Uhlenberg (2005) suggested that age-segregated institutions (e.g., schools, colleges, care homes, and retirement villages) are the root of ageism. These segregated community spaces imply vulnerability of both old and young and act to segregate older people from their wider communities. It is important to examine such structures and the norms that they shape, as well as to examine the lived experiences of the individuals who reside in them.

Other less tangible institutions play an equally significant role in circulating dominant representations of ageing. The media and popular culture offer a powerful outlet for stereotypical images of ageing. The dominant social representation of old age as passive and as invisible is fuelled by mainstream media, film, and television that have historically heralded youth as normal and aged as 'other'. Mass media are able to utilise objectification, a powerful means through which social representations come into circulation. This is the process by which images or symbols are used to make an abstract idea more concrete. This might involve pictures of complex ideas being personified or anthropomorphised as in the example of 'Dolly', the first sheep to be successfully cloned, a personification of a complex unfamiliar process (Bauer and Gaskell 1999). Many complex or unfamiliar and new phenomena are first objectified by the media (Flick 1998; Hoijer 2011), resulting in societal images and stereotypes which demonise marginalised or poorly represented groups. Objectification is a process that simplifies and reduces complexities and diverse ideas such as what it means to be aged to simpler images such as that of a fragile older woman in a care home bed. When persistent and widespread, these images can come to dominate what we understand old age to be.

Old as Other

The objectification of old age through images that highlight crude distinctions from youth or extreme case examples fuel notions of old as 'other'. Othering, a process linked to identity and said to serve a protective function, draws on fears and the thema (i.e., binary) of 'them' and 'us' (Smith et al. 2015). This us/them distinction fuels and fosters prejudice. It highlights and emphasises differences and plays on vulnerabilities and deep-rooted fears. Ageism, unlike racism, sexism, and homophobia where prejudice is directed against relatively fixed social categories (whether socially constructed or biologically determined), is a prejudice towards a much more transient social category. We all begin as young and if we are lucky, get to grow old. Age-related frailty and dementia are, however, a source of fear for many as well as being a major policy concern in many parts of the world (Higgs and Gilleard 2017). Smith et al. (2015) suggest that our sense of self is protected somewhat and is held safe by representing old as other, by distancing ourselves and associating death, decline, and fragility with something very different and distinct from ourselves. Many refer to ageism as the 'fear of our future selves' (Calasanti 2008). Those in positions of power and influence participate in othering (Joffe 2007). This may be consciously, as a means of shifting blame from one group to another, or unconsciously as many social representations are rooted in fears and emotions. Through language and actions, powerful social actors can feed negative social representations. This othering can be seen as both an exploitation of fears of ageing as well as a reaction to it.

Othering often has its basis in themata. Themata are simplistic binary oppositions or antinomies such as self/other, life/death, and young/old that are deeply rooted in culture (Markova 2003). Moscovici (1992) proposed that all common-sense thinking and therefore all social representations have their basis in themata (e.g., as us/them). These pairs of antinomies or themata serve to help make sense of issues in the social world (Smith et al. 2015). Themata are evident in language and in action though are often less explicit than the social representations they underpin (Markova 2003; Moloney et al. 2015). A single social representation may be underpinned by several themata, or a single thema may be at the root of a diverse system of social representations (Smith et al. 2015). The young/old distinction is one example of a thema that has helped legitimise dominant negative social representations of ageing.

The Reach of Dominant Social Representations

This legitimisation of social representations of ageing as unwanted and negative is achieved through objectification, othering, themata, and other conscious and unconscious social mechanisms. These social representations are reflected and represented in everyday communication and action, in turn perpetuating the dominance of particular ideas and images of ageing. Negative social representations serve to exclude or limit opportunities for older people (Carpentier et al. 2008). However, the malleable nature of a social representation can be exploited in attempts to promote positive change. Some studies have demonstrated how the creation of environments with an explicit focus on more positive social representations of ageing can result in tangible improvements to the lives of individual older adults. Dryjanska et al. (2017) demonstrated how an emphasis on positive social representations within the production of a theatrical performance led to an improvement in older adults' memory when compared to a control group. Such research illustrates the interplay between positive and negative social representations and individual health and wellbeing in a tangible, measurable way. Social, cultural, and physical environments in which negative social representations of ageing dominate through the language and actions of those with greater power have consequences for those older adults who interact with those environments. Resistance and positive social change are, however, possibilities, as there is also agency and malleability in social representations. Negative social representations can be challenged and resisted through empowerment, through positive individual and community narratives, and through collective action which are considered in the remainder of this chapter.

Narratives of Ageing

At the core of social representations of ageing in Western society is a dominant cultural narrative of ageing as being a process of inevitable decline in physical and mental capacity. This cultural narrative is pervasive and reflected in the media and our social institutions. It is for this reason that Gullette (2003) has argued that people do not simply age but are aged by culture. However, Freeman (2008) has suggested that behind this dominant narrative is a counter-narrative of the 'vital, self-sufficient individual' (p. 175). It is also one that is associated with wisdom and fortitude rather than recklessness and impulsivity. Older people develop and announce their identity through a process of narrative engagement with these cultural narratives of old age (Hammack and Pilecki 2012). This process of engagement can be both as an individual and as part of a collective. Here we consider two reactions to the cultural narrative of decline explored by Freeman (2011) and then consider some social and political strategies of resistance.

Narrative Foreclosure

Freeman (2011) described acceptance of the cultural narrative of ageing as decline as narrative foreclosure, the reaction of those older people who feel that their future life story is foreclosed. It is the 'premature conviction that one's life has effectively ended' (Freeman 2010, p. 83). For these people, the dominant narrative of decline is accepted as inevitable with little prospect of development; the future is one with few prospects for growth but only the expectation of further decline. However, such narrative foreclosure is not absolute but will vary depending upon how the individual interprets both past events and future possibilities.

Freeman distinguished between four different forms of narrative foreclosure. The first he termed 'dead ends'. This is the case when there is perceived to be no prospect of new opportunities available. For this person, it 'was too late for opening up a new chapter' (p. 5). There was little to do except wait for the end which was inevitable. This person could be described as being in the final chapter of life or in the waiting room for the ending. The second form Freeman termed 'the point of no return', which was the awareness that you could not go back to the past but that there was still the prospect of new opportunities. The challenge was to develop these new opportunities rather than to focus on past achievements. The third form of foreclosure was a growing awareness that past negative events were over and did not need to determine future events. To illustrate this process, Freeman considered the case of Primo Levi, a survivor of the Nazi concentration camps. Levi could look back on his past life then as a period of living on an animal level. Despite the shame about that previous experience, there was escape through an awareness of it only having lasted for a period and thus there were now growth possibilities. It was possible to 'reopen his life and restart the creative process' (p. 7). The extent to which this narrative leap is possible will depend upon how the person engages with past traumatic events. A frequent strategy is to ignore these past events and focus on developing a new narrative. In Levi's case, this was not possible and he tragically committed suicide. For others, developing a story to capture past negative experiences can be a life-long quest inhibiting the possibility of moving forward (see Brooks 1994).

The final fourth form of narrative foreclosure considered by Freeman was foreclosure as existential despair. To illustrate this process, he considered the case of Ivan Ilyich, the character in Tolstoy's famous novella *The Death of Ivan Ilyich*. Tolstoy described the life of the magistrate Ivan Ilyich who had a successful career but then fell and is diagnosed with a terminal condition. Reflecting on his life, he begins to see it as filled with certain successes but that these are marked by self-interest and now appear to him as basically worthless. Tolstoy contrasts this perceived wasteful life with an authentic life which is full of compassion and concern for others. But then, in this moment of existential despair there is light for Ivan Ilyich—the realisation that he can do something right now. Tolstoy details this process in almost religious terms: 'and it was revealed to him that though his life was not what it should have been, this could still be rectified' (Tolstoy 1886/1960, p. 151). By showing love and compassion for his family, Ivan Ilyich could develop a new authentic narrative. This transformation requires him to accept but separate himself from his negative life story which was the past and develop a new positive narrative.

Admittedly, these four forms of narrative closure do not encompass the reaction of all or indeed most older people. There remains the cultural narrative of the vital and dynamic older person repressed by social institutions which equate youth with vitality. Engaging with this cultural narrative offers the prospect of new opportunities.

Narrative Openness

Freeman (2010) suggests that the opposite of narrative foreclosure is narrative openness or narrative freedom. By this he meant the awareness that despite the many challenges of ageing, life still provides many opportunities. The dominant life narrative that had shaped earlier activities, including what were perceived as burdensome social responsibilities, have now been removed. Instead the older person's narrative is one of developing new activities or indeed activities that had been curtailed because of occupational and domestic demands (see also Chap. 2, this volume). Consider the case of the older person who in their youth aspired to be an artist but instead ended up in a more routine occupation with little opportunity for creativity. Now that they have retired, they have the opportunity of once again taking up that earlier ambition or alternatively developing new ambitions. The release from everyday demands provides the opportunity to develop new narrative scripts.

Trethewey (2001) in her study of the midlife women noted that many of those that she interviewed highlighted the benefits of growing older. They were not frustrated by the passage of youth but welcomed the onset of old age as providing a sense of maturity and release from the professional rat race. Some women welcomed the loss of sexual identity as it lessened the unwelcome forms of attention and sexual harassment that they had experienced in the workplace and elsewhere. She noted that for some, the loss of sexuality entailed 'losing the burdensome sexual objectification, and sexual tension that they too often experienced as younger professionals' (p. 200). As they grew older they were freed of these expectations and felt less need to adopt a defensive orientation towards others.

Such narrative freedom can take various forms from the more mundane to the more dynamic. At the more mundane level is the contentment which flows from acceptance of old age. Instead of looking to the future with frustration, the narrative is about acceptance of the ending of one [working] life and the potential opening up of another. In the play 'Waiting for Godot', Samuel Beckett (1956) has this exchange between the characters Vladimir and Estragon who are discussing the imagined voices of the dead which they compare to the rustling of leaves:

VLADIMIR: What do they say?

ESTRAGON: They talk about their lives.

VLADIMIR: To have lived is not enough for them.

ESTRAGON: They have to talk about it.

This exchange about talking applies to older people reasserting their existence as living social beings. They may not be developing new opportunities but they still have lots to talk about. They are living life in the now through their very conversation. For old people such conversation can be the source of contentment, but it is one which they unfortunately do not always have access to. AgeUK (2017) has estimated that 3.6 million older people in the UK live alone and 1.9 million report feeling ignored and isolated. It was for this reason that the Jo Cox Commission on Loneliness (2017) titled their report *Combatting loneliness one conversation at a time. A call to action.* In these conversations are the small stories that bind people together and reaffirm their identities (Bamberg 2006).

Another alternative to acceptance of decline being perceived in a negative light is the adoption of what Randall (2013) has described as an ironic narrative stance to our lives as individuals and towards life in general. He defines irony as 'the acceptance (if not embracing) of ambiguity and relativity' (p. 166). If earnestness is typical of youth, irony more reflects old age. It is the period when we become more accepting of the multiple interpretations possible of our lives. Even more, this acceptance can lead to a certain contentment rather than frustration at ageing at the physical, social, cognitive, and spiritual level. At the physical level, an ironic stance can reflect a certain wry amusement at decline in our physical capacity. At the social level, an ironic stance reflects a certain contentment with disengagement from the social world. At the cognitive level, an ironic stance can enable us to accept multiple perspectives rather than searching for the right answer. Finally, at the spiritual level, an ironic stance reflects an awareness of our mortality and a certain gerotranscendence. Randall suggests that the uses of such strategies as guided autobiography, creative reminiscence, and life-writing with different literary forms are potential approaches to promoting forms of ironic narrative.

Social and Political Narratives: Ageing as Progress

For older people, becoming old is not a singular event but a process within which our sense of narrative identity evolves through ongoing resistance to the dominant cultural narrative of decline. According to Gullette (2003), adulthood is not the pinnacle of achievement but rather a step on a continuum that has many twists and turns. There are many opportunities for development in later years. She terms this process the progress narrative. It is something to be worked on rather than something that is inevitable. Let us consider some strategies for developing a progress narrative as we grow older.

Everyday Life Resistance to negative ageist stereotypes can be part of everyday life. This process is not confined to older people but can start in middle age as often people react strongly to being categorised as old. This can be exhibited in the style of clothing and participation in certain activities. For instance, in our study of middle-aged baby boomers' social representations of health and ageing we found that these people frequently expressed resentment at any indication that they were old, for example, being called sir or madam (Murray et al. 2004). This baby-boomer generation had internalised the desire 'to die before I get old' and wanted to be 'forever young'. In their study of anti-ageing strategies among middle-aged people, Calasanti et al. (2016) found that although the use of anti-ageing cosmetic products was widespread, the users referred to health and bodily appearance rather than anti-ageing as a justification for their usage. For them, ageing was not something to be combatted rather their current lives should be enhanced. Their life story was one of progress.

Continued Work The removal of the formal UK retirement age has provided many people with the opportunity to continue working after the statutory pension age (SPA). In a survey of older workers, Smeaton et al. (2009) found that many continued working not simply to maintain their income but also because they enjoyed their work, the social contact, and the routine. There is evidence that the urge to delay retirement is more pronounced among men who have greater attachment to their worker identity (Barnes and Parry 2004). In an interview study of a sample of older people who continued to work after SPA, Reynolds et al. (2012) found that work provided not only increased financial security but also a sense of control over their health and an opportunity for personal development. This latter theme included the various components of life as progress including learning, challenge, and resistance to ageist stereotypes. They resented the expectation that now that they were retired, they should engage in activities such as golfing and cruises. Admittedly, the study participants tended to be middle class for whom work was often very enjoyable. For others, retirement can provide the opportunity to escape from monotonous routines and to develop new exciting opportunities.

After retirement, many older people become involved in various forms of volunteer activity. Such activity provides many of the benefits of paid employment including social contact and sense of purpose. In a study of the role of older volunteers in a theatre, Bernard et al. (2015) found that they emphasised the opportunity it provided to challenge the dominant negative representation and to develop new skills. For example, one older volunteer commented:

> I think that as you do get older, if you don't have a manageable challenge … you're not developing yourself anymore, and I think that is one of the things that the theatre does, not just for the young people but for us. It's still developing us, and that's what's so great. (p. 100)

Again, the emphasis was on developing a progress narrative.

Research Focus on Arts Activities

A series of 'New Dynamics of Ageing' projects demonstrated the power of arts participation to develop a counter-narrative of ageing (Murray et al. 2014). Murray and Crummett (2014) explored the use of community arts as a means of promoting increased social engagement among older residents of a disadvantaged urban neighbourhood. The experience of the older residents was one of social exclusion based on age and social class. Discussion with a group confirmed their frustration at what they perceived as a very negative outsider representation of their neighbourhood. They talked about local council officials as having negative expectations of local residents.

Participation in a community arts project provided them with an opportunity to develop a new community narrative of talent, which both challenged the talentless self-narrative and the negative outsider representation. Extracts from conversations with participants illustrate this process (Murray and Crummett 2010). One of the male participants enjoyed the silk screen printing and commented afterwards:

> I could do what I wanted. I didn't have to constrain. I could experiment. All my work is the way I want to do it. … I didn't know what to expect because we've never had this before so I wasn't sure what to expect. I was surprised that we had so much talent. I mean there are certain people that have astounded me. I mean Mary, her work has been very good, and Monica surprises me all the time.

Another female participant reflected on the opportunity the art provided for her to escape from everyday concerns:

> I used to think well I've got no time to take for me but now with this, it is time for me. You get lost in what you're doing and you've no time to think what's happening around you. Well, we're just here doing your project and just for a couple of hours you're in your own world doing what you want and nobody's there to interfere with it.

An important part of the artwork was the final product that the participants could admire and show to others. They talked about each other's work and supported each other. As the same male participant said: 'One of the main points is you can turn round or look at whatever it is and think and yes, I made that.' The artwork was a concrete expression of their abilities that they could talk about with others in the project or in the wider community. One of the women felt that other members of the community also had many abilities but they did not express them because 'they're a bit frightened, they think – I can't do that'.

However, there was also a broader social challenge. The participants felt that they were able to demonstrate to outside officials and others their talents that they felt had not been recognised. In publicly displaying their artwork at the end of each project, they were challenging the perceived negative social representation of their community as being talentless. An important part of the community artwork was its participatory nature. The older people resented being told what to do. For example, at one early meeting of the project some council officials detailed their expectations. Although this was done in a low-key fashion, the older people felt uncomfortable. One female participant subsequently told one in the research team that she felt she was being told what to do and she added:

> I'm really angry and so are other people. [...] They wanted to tell us how we had to do things. [...] Well they can stuff it. I'm going to get my big table out and we'll meet in my back kitchen. We don't want those people telling us what to do.

The older residents enjoyed making art together, but they did not want to be told what to do. This brought back to many of them memories of officialdom and curtailment of their activities. In making art, they challenged those traditional restrictions and showed officialdom what they could do on their own when they had the opportunity. They were beginning to develop a new progress narrative of talent and potential.

In the 'Representing self – representing ageing' project (Hogan and Warren 2012), older women worked to critique dominant visual narratives of ageing and to develop their own images of ageing. They did this through participating in arts workshops and using a range of techniques to create alternatives images which were displayed in an exhibition. These collaborative arts activities developed among the older women alternative narratives of creativity and solidarity.

In the Fiction and the Cultural Mediation of Ageing Project (Hubble and Tew 2013) older people participated in a group reading project in which they collectively critiqued the representations of older people in a selection of popular novels. The older people accepted that the more negative representations may have been typical many years ago, but they did not apply today. They argued that many of the novelists they discussed were living in the past. In addition, the older people in the project not only discussed the novels but also developed counter stories in their diaries. In a related project de Medeiros (2014) in her study of creative writing groups for older people explored encouraging the participants to experiment with alternative forms of writing. She argued that in later life our expectations are shaped by the dominant cultural narratives of the journey, the family, and decline. Using the standard autobiographical form of writing tends to reinforce this dominant narrative. In groups, she encouraged participants to challenge this narrative and to co-create alternatives forms as a way of developing a new counter story.

In another related project, Seymour and Murray (2016) considered the impact of shared poetry reading by older residents of an assisted living facility. In this study, the opportunity to participate in the shared reading challenged any narrative foreclosure and instead encouraged the participants to develop new ways of seeing the world. One of the participants reflecting on her life articulated a clear process narrative:

> I think we renew ourselves in the way we think. … Like when we retire we got to sort of learn … you know, when you're married you've got your partner and your family. And when you're on your own you have to start a new way of thinking. There's a renewal because you can't live like you did because things have changed. (p. 197)

Intergenerational Action A community strategy for combatting the negative cultural narrative of ageing is intergenerational action. Sometimes this is organic and involves grassroots interaction of young and old in a community. However, such action can also occur as more structured 'intergenerational practice' facilitated by community development workers or charitable organisations. Intergenerational practice aims to provide older and younger people with mutually beneficial activities and opportunities to foster community cohesion. For older people, the involvement of younger people in social change efforts can provide community allies, added energy and support in the long-term sustainability of an initiative (Statham 2009). However, there is a danger that intergenerational work, when facilitated unreflexively can indeed perpetuate negative images of ageing through the involvement of exclusionary practices or limited opportunities. Therefore, mutuality, support, and inclusion are vital to the success of intergenerational practice.

In the UK, 'Magic Me' (https://magicme.co.uk/) is one of the largest providers of intergenerational arts projects for older people across London and beyond. By engaging both older and younger generations in arts-based activities, both groups are able to learn new skills and be more involved in the wider community. The organisation embraces an ethos of inclusivity and positive social change and trains others (e.g., community workers, service providers) across the country to work with older and younger communities in this way. A commitment to inclusive, empowering practice by those who work with older communities can work to resist and challenge the more demoralising and exclusionary practices and in doing so challenge what it means to be old. 'Linking Generations Northern Ireland' (http://linkinggenerationsni.com/) offer a similar example of intergenerational working being facilitated not only at a small group level but also through work with community leaders, developers, and service providers. Working with those in positions of power to provoke critical reflection on practice is another means of resisting negative dominant social representations and tackle the age-segregated nature of many social institutions.

Social and Political Action Moving beyond forms of individual and community action requires challenging the dominant political narrative which excludes older people from participation in civic life. Toolis and Hammack (2015) in their study of homelessness argue that dominant narratives of safety and economic vitality construct the boundaries of exclusion of homeless people. These narratives justify the criminalisation of homelessness. Similarly, dominant narratives of economic vitality construct boundaries of exclusion of older people from civic life. Instead they are portrayed as not contributing to economic life and as such not deserving of the various benefits. Challenging this narrative requires political action at different levels. At the policy level there is the move towards the promotion of age-friendly cities and communities (Buffel and Handler 2017). This can take many forms from involvement of older people in the planning of housing and facilities through to the provision of services that meet the particular needs of older people.

Some older people become actively involved in various forms of social and political action beyond community activities. This can range from volunteering for voluntary organisations to more obvious political action. In their study of those who participated in older people's forums, Barnes et al. (2012) found that for many it was a form of resistance to being old. Through their actions, the volunteers were promoting a counter-narrative of action.

Other older people continue to or become involved in various forms of political action. Molly Andrews has conducted extensive research on the experiences of political activists. This has involved detailed interviews with activists in England, the USA, the former East Germany, and South Africa (Andrews 1991, 2007). In their life stories, the older activists interwove the personal and the social narratives—they were part of a broader political movement. One feature that Andrews noted in her conversations was the reluctance of the older activists to focus on themselves in their recollection of past activities. Instead they emphasised that their generation had forsworn personal interest for involvement in a broader struggle for social justice. This was something which did not end as they grew older; rather it was part of their lives. As one of her participants said: 'You must not give up, you must keep pegging away' (Andrews 2007, p. 66).

For these older activists, participation in the political struggle was an integral part of their narrative identity. To cease involvement would be to disrupt that identity. Eileen, one of those interviewed by Andrews (2007) put it this way: 'It gives you a motive for going on living. It is very strong. It's survival' (p. 67). She was echoing the words of the English novelist and social activist Edward Upward (1903–2009) who titled the final volume of his semi-autobiography trilogy 'No home but the struggle' (2007), which conveyed the continuing importance of the political struggle for social justice in his identity as he grew older.

In their study of female activists, Chazan (2016) conducted interviews with 32 older women from which they compiled an archive of 'activist herstories'. Their analysis of these herstories revealed four central motivations for social activism: (1) time and space: the release from family and work responsibilities; (2) perspective: the opportunity to stand back and take a broader perspective on life and realise that there is no time to lose;

(3) grand motherhood: whether or not they had grandchildren they felt a social responsibility for the next generation; and (4) combatting ageism. Each of these provided an opportunity for developing a progressive political narrative.

These examples are of older people who had a lifetime involvement in forms of political action. Some older people are new to political action as in their earlier lives they had not been so involved. This was illustrated in a study by Guillemot and Price (2017) who interviewed some visitors to a community centre that was threatened with closure. These older people organised a protest outside the local town hall. When interviewed, these older activists emphasised that although they had never been involved in such protest before they felt that this was necessary because of the importance of the community centre to them. As one of the protestors said:

> I think this [the day centre] is a lifesaver for me, for all these people here. It can't close, it can't close. We'd all sit at home and fade away. As I said, I'm not doing it yet. I'm not going to fade away yet. (p. 59)

Analysis of the interviews identified two key features of the protest: the perceived threat (closure of centre) and the role of the staff in supporting the protest—a form of proxy efficacy (Bandura 2000). In addition, an important feature of the protest was the extent to which it was perceived as a public shaming of the authorities. To enhance this shaming some of the older people referred to the importance of attending the protest in their wheelchairs. As another one of the interviewees said, 'I think to see a few people in wheelchairs, it does help' (Guillemot and Price 2017, p. 61). Through circumstance they were developing a new narrative identity of social activists.

A final example of an emerging progressive political narrative is the work of the Raging Grannies. This is a social movement of older women which emerged from the Canadian peace movement in the 1980s. It has spread to over 100 groups (or gaggles) across North America and elsewhere. These groups engage in a range of activities to promote social and environmental justice including protests against war, climate change, poverty, for refugee and minority rights, and so on. Their modus operandum has three central features: dressing in flamboyant costumes, singing satirical songs, and going to places where they are not invited. Chazan (2016) in her study of a group of these women found that their aims were to challenge the cultural narrative that they did not have a role to play in political change and to challenge the idea that social protest is only the purview of younger people. Participation in these activist groups was a means of resisting these stereotypes. These raging grannies were redefining old age in positive terms through growing old disgracefully (Caissie 2011).

Personal Stories Are Political

Finally, we can consider the power of personal stories to challenge dominant cultural narratives. As discussed before (Murray 2013), storytelling is a powerful means of mobilising support for minority ideas. Poletta (2006) has described the power of personal stories to convince others of the legitimacy of arguments—'to chip away at the wall of public indifference' (p. 2). A recent graphic example of the power of publicly telling personal stories was the abortion referendum in Ireland. This was detailed in a commentary by Fintan O'Toole (2018):

> the old feminist slogan that the personal is political holds true, but it also works the other way around. The political has to be personalised. The greatest human immune system against the viruses of hysteria, hatred and lies is storytelling. Even when we don't trust politicians or experts, we trust people telling their own tales. We trust ourselves to judge whether they are lying or being truthful. Irish women had to go out and tell their own stories, to make the painful and intimate into public property.

In the case of ageing, older people can publicly announce their achievements and their defeats as a means of building collective resistance to dominant cultural narratives of decline. The future is not foreclosed or simply more of the same but a place for forms of activity which can disrupt negative social representations of ageing and open up new opportunities for older people.

Conclusion

In this chapter, we have considered the oppressive role of dominant negative social representations of ageing and various strategies of resistance ranging from the personal to the political. However, it is in the transition from the personal to the political that we want to highlight. While in everyday life we may attempt as individuals to challenge the negative social representation of ageing, this places responsibility for this process on the individual. While people that are middle class may have more resources to develop a resistant narrative at a personal level, there is a need to develop more opportunities for developing various forms of collective resistance for more, older people. Nolas et al. (2017) have emphasised that resistance can take many forms. As they stated: 'political activism can be found in mundane activities as well as in banal spaces' (p. 9). It is not confined to those who have previously been resistant but can develop among large groups of older people who have previously been quiescent and accepting. Indeed, many older people are keen to become involved in various forms of social and political action, the fruits of which they may not live to see (Andrews 2017). Enhancing the lives of older people involves facilitating more opportunities for their involvement in such action. Community and political psychology strives to understand the processes that facilitate such involvement.

References

AgeUK. (2017). *Loneliness*. https://www.ageuk.org.uk/information-advice/health-wellbeing/loneliness/

Andrews, M. (1991). *Lifetimes of Commitment: Ageing, Politics, Psychology*. Cambridge, UK: CUP Archive.

Andrews, M. (2007). *Shaping History. Narratives of Political Change*. Cambridge: Cambridge University Press.

Andrews, M. (2017). Enduring Ideals: Revisiting *Lifetimes of Commitment* Twenty-five Years Later. *Contemporary Social Science, 12*(1–2), 1–12.

Bamberg, M. (2006). Stories: Big or Small? Why Do We Care? *Narrative Inquiry, 16*, 147–155.

Bandura, A. (2000). Exercise of Human Agency Through Collective Efficacy. *Current Directions in Psychological Science, 9*, 75–78.

Barnes, H., & Parry, J. (2004). Renegotiating Identity and Relationships: Men and Women's Adjustments to Retirement. *Ageing & Society, 24*(2), 213–233.

Barnes, M., Harrison, E., & Murray, L. (2012). Ageing Activist: Who Gets Involved in Older People's Forums? *Ageing & Society, 32*, 261–280.

Bauer, M. W., & Gaskell, G. (1999). Towards a Paradigm for Research on Social Representations. *Journal for the Theory of Social Behaviour, 29*(2), 163–186.

Brooks, P. (1994). *Psychoanalysis and Storytelling*. Hoboken: Blackwell.

Buffel, T., & Handler, S. (2017). *Age-Friendly Cities and Communities: A Global Perspective*. Bristol: Policy.

Buffel, T., Phillipson, C., & Scharf, T. (2012). Ageing in Urban Environments: Developing 'age-friendly' Cities. *Critical Social Policy, 32*(4), 597–617.

Caissie, L. (2011). The Raging Grannies: Narrative Construction of Gender and Ageing. In G. Kenyon, E. Bohlmeijer, & W. L. Randall (Eds.), *Storying Later Life* (pp. 126–142). Oxford: Oxford University Press.

Calasanti, T. (2008). A feminist confronts ageism. *Journal of Aging Studies, 22*(2), 152–157.

Calasanti, T., King, N., Pietilä, I., & Ojala, H. (2016). Rationales for Anti-ageing Activities in Middle Age: Ageing, Health, or Appearance. *The Gerontologist, 58*(2), 233–241.

Carpentier, N., Ducharme, F., Kergoat, M. J., & Bergman, H. (2008). Social Representations of Barriers to Care Early in the Careers of Caregivers of Persons with Alzheimer's Disease. *Research on Aging, 30*(3), 334–357.

Chazan, M. (2016). Settler Solidarities as Praxis: Understanding 'Granny Activism' Beyond the Highly Visible. *Social Movement Studies, 15*(5), 457–470.

Dryjanska, L., Aiello, S., & Giua, M. (2017). Social Representations, Ageing and Memory: A Holistic Approach to Cognitive Assessment. *Ageing & Society, 37*(4), 804–822.

Elcheroth, G., Doise, W., & Reicher, S. (2011). On the Knowledge of Politics and the Politics of Knowledge: How a Social Representations Approach Helps Us Rethink the Subject of Political Psychology. *Political Psychology, 32*(5), 729–758.

Flick, U. (1998). The Social Construction of Individual and Public Health: Contributions of Social Representations Theory to a Social Science of Health. *Social Science Information, 37*(4), 639–662.

Flick, U., Fischer, C., Neuber, A., Schwartz, F. W., & Walter, U. (2003). Health in the Context of Growing Old: Social Representations of Health. *Journal of Health Psychology, 8*(5), 539–556.

Freeman, M. (2008). Beyond Narrative: Dementia's Tragic Promise. In *Health, Illness and Culture* (pp. 175–190). Abingdon: Routledge.
Freeman, M. (2010). *Hindsight. The Promise and Peril of Looking Backward.* New York: Oxford University Press.
Freeman, M. (2011). Narrative Foreclosure in Later Life: Possibilities and Limits. In G. Kenyon, E. Bohlmeijer, & W. L. Randall (Eds.), *Storying Later Life* (pp. 3–19). Oxford: Oxford University Press.
Guillemot, J. R., & Price, D. J. (2017). Politicisation in Later Life: Experience and Motivations of Older People Participating in a Protest for the First Time. *Contemporary Social Science, 12*(1–2), 52–67.
Gullette, M. M. (2003). From Life Storytelling to Age Autobiography. *Journal of Aging Studies, 17*, 101–111.
Hagestad, G. O., & Uhlenberg, P. (2005). The Social Separation of Old and Young: A Root of Ageism. *Journal of Social Issues, 61*(2), 343–360.
Hammack, P. L., & Pilecki, A. (2012). Narrative as a Root Metaphor for Political Psychology. *Political Psychology, 33*(*1*), 75–103.
Herzlich, C. (1974). *Health and Illness: A Social Psychological Analysis*. London: Academic.
Higgs, P., & Gilleard, C. (2017). Ageing, Dementia and the Social Mind: Past, Present and Future Perspectives. *Sociology of Health & Illness, 39*(2), 175–181.
Hogan, S., & Warren, L. (2012). Dealing with Complexity in Research Processes and Findings: How Do Older Women Negotiate and Challenge Images of Ageing? *Journal of Women and Aging, 24*(4), 329–350.
Höijer, B. (2011). Social Representations Theory. *Nordicom Review, 32*(2), 3–16.
Howarth, C. (2006). A Social Representation Is Not a Quiet Thing: Exploring the Critical Potential of Social Representations Theory. *British Journal of Social Psychology, 45*(1), 65–86.
Howarth, C., Campbell, C., Cornish, F., Franks, B., Garcia-Lorenzo, L., Gillespie, A., et al. (2013). Insights from Societal Psychology: The Contextual Politics of Change. *Journal of Social and Political Psychology, 1*(1), 364–384.
Howarth, C., Andreouli, E., & Kessi, S. (2014). Social Representations and the Politics of Participation. In P. Nesbitt-Larking, C. Kinnvall, & T. Capelos (Eds.), *The Palgrave Handbook of Global Political Psychology* (pp. 19–38). London: Palgrave Macmillan.
Hubble, N., & Tew, P. (2013). *Ageing, Narrative and Identity: New Qualitative Social Research*. London: Palgrave.
Jo Cox Commission on Loneliness. (2017). *Combatting Loneliness One Conversation at a Time. A Call to Action*. https://www.jocoxloneliness.org/

Joffe, H. (1999). *Risk and 'the Other'*. Cambridge, UK: Cambridge University Press.
Joffe, H. (2007). Anxiety, Mass Crisis and 'the Other'. In *Public Emotions* (pp. 105–118). London: Palgrave Macmillan.
Jovchelovitch, S. (2007). *Knowledge in Context: Representations, Community and Culture*. London: Routledge.
Kuhn, T. S. (1970). *The Structure of Scientific Revolutions*. Chicago: Univ. of Chicago Press.
Marková, I. (2003). *Dialogicality and Social Representations: The Dynamics of Mind*. Cambridge: Cambridge University Press.
McCann Mortimer, P., Ward, L., & Winefield, H. (2008). Successful Ageing by Whose Definition? Views of Older Spiritually Affiliated Women. *Australasian Journal on Ageing, 27*(4), 200–204.
Moloney, G., Gamble, M., Hayman, J., & Smith, G. (2015). Without Anchor: Themata and Blood Donation. *Papers on Social Representations, 24*(2), 1–21.
Morant, N. (2006). Social Representations and Professional Knowledge: The Representation of Mental Illness Among Mental Health Practitioners. *British Journal of Social Psychology, 45*(4), 817–838.
Moreno, X., Sánchez, H., Huerta, M., Albala, C., & Márquez, C. (2016). Social Representations of Older Adults Among Chilean Elders of Three Cities with Different Historical and Sociodemographic Background. *Journal of Cross-Cultural Gerontology, 31*(2), 115–128.
Moscovici, S. (1992). The Psychology of Scientific Myths. In M. Von Cranach & E. M. Lipiansky (Eds.), *Social Representations and the Social Bases of Knowledge* (pp. 3–9). Leviston: Hogrefe & Huber.
Moscovici, S. (1988). Notes Towards a Description of Social Representations. *European Journal of Social Psychology, 18*(3), 211–250.
Moscovici, S. (2008 [1976]). *Psychoanalysis: Its Image and Its Public*. Cambridge: Polity
Moscovici, S., & Marková, I. (1998). Presenting Social Representations: A Conversation. *Culture & Psychology, 4*(3), 371–410.
Murray, M. (2002). Connecting Narrative and Social Representation Theory. *Social Science Information, 41*, 653–673.
Murray, M. (2013). Implementation: Putting Analyses into Practice. In U. Flick (Ed.), *The Sage Handbook of Qualitative Data Analysis* (pp. 585–599). London: Sage.
Murray, M., & Crummett, A. (2010). [Conversations with Young at Heart Participants]. Unpublished raw data.

Murray, M., & Crummett, A. (2014). I Don't Think They Knew We Could Do These Sorts of Things': Social Representations of Community and Participation in Community Arts by Older People. *Journal of Health Psychology, 15*, 777–785.

Murray, M., Pullman, D., & Rodgers, T. H. (2003). Social Representations of Health and Illness Among 'Baby–Boomers' in Eastern Canada. *Journal of Health Psychology, 8*(5), 485–499.

Murray, M., Nelson, G., Poland, B., Maticka-Tyndale, E., & Ferris, L. (2004). Assumptions and Values of Community Health Psychology. *Journal of Health Psychology, 9*(2), 323–333.

Murray, M., Amigoni, D., Bernard, M., Crummett, A., Goulding, A., Munro, L., Newman, A., Rezzano, J., Rickett, M., Tew, P., & Warren, L. (2014). Understanding and Transforming Ageing Through the Arts. In A. Walker (Ed.), *The New Science of Ageing* (pp. 77–112). Bristol: Policy Press.

Nolas, S.-M., Varvantakis, C., & Aruldoss, V. (2017). Political Activism Across the Life Course. *Contemporary Social Science, 12*(1–2), 1–12.

O'Connor, E. (2006). Glassblowing Tools: Extending the Body Towards Practical Knowledge and Informing a Social World. *Qualitative Sociology, 29*(2), 177–193.

O'Toole, F. (2018). If Only Brexit Had Been Run Like Ireland's Referendum. *The Guardian*. Retrieved May 30, 2018, from www.theguardian.com

Poletta, F. (2006). *It Was Like a Fever: Storytelling in Protest and Politics*. Chicago: University of Chicago Press.

Quéniart, A., & Charpentier, M. (2012). Older Women and Their Representations of Old Age: A Qualitative Analysis. *Ageing & Society, 32*(6), 983–1007.

Randall, W. L. (2013). Aging, Irony and Wisdom: On the Narrative Psychology of Later Life. *Theory & Psychology, 23*(2), 164–183.

Renedo, A., & Jovchelovitch, S. (2007). Expert Knowledge, Cognitive Polyphasia and Health: A Study on Social Representations of Homelessness Among Professionals Working in the Voluntary Sector in London. *Journal of Health Psychology, 12*(5), 779–790.

Reynolds, F., Farrow, A., & Blank, A. (2012). 'Otherwise It Would Be Nothing But Cruises': Exploring the Subjective Benefits of Working Beyond 65. *International Journal of Ageing and Later Life, 7*(1), 79–106.

Seymour, R., & Murray, M. (2016). When I Am Old, I Shall Wear Purple: A Qualitative Study of the Effect of Group Poetry Sessions on the Wellbeing of Older Adults. *Working with Older People, 20*(4), 195–198.

Smeaton, D., Vegeris, S., & Sahin Dikmen, M. (2009). *Older Workers: Employment Preferences, Barriers and Solutions* (Research Report 43). Manchester: Policy Studies Institute.

Smith, N., O'Connor, C., & Joffe, H. (2015). Social Representations of Threatening Phenomena: The Self-Other Thema and Identity Protection. *Papers on Social Representations, 24*(2), 1–23.

Statham, E. (2009). Promoting Intergenerational Programmes: Where Is the Evidence to Inform Policy and Practice? *Evidence & Policy: A Journal of Research, Debate and Practice, 5*(4), 471–488.

Toolis, E. E., & Hammack, P. L. (2015). 'This Is My Community': Reproducing and Resisting Boundaries of Exclusion in Contested Public Spaces. *American Journal of Community Psychology, 56*(3–4), 368–382.

Trethewey, A. (2001). Reproducing and Resisting the Master Narrative of Decline. Midlife Professional Women's Experiences of Aging. *Management Communication Quarterly, 15*(2), 183–226.

Twigg, J. (2004). The Body, Gender, and Age: Feminist Insights in Social Gerontology. *Journal of Aging Studies, 18*(1), 59–73.

Index

E. Peel et al. (eds.), *Psychologies of Ageing*,
https://doi.org/10.1007/978-3-319-97034-9